Maintenance, Reliability and Troubleshooting in Rotating Machinery

Scrivener Publishing
100 Cummings Center, Suite 541J
Beverly, MA 01915-6106

Publishers at Scrivener
Martin Scrivener (martin@scrivenerpublishing.com)
Phillip Carmical (pcarmical@scrivenerpublishing.com)

Maintenance, Reliability and Troubleshooting in Rotating Machinery

Volume 2

Edited by
Robert X. Perez

Scrivener Publishing

WILEY

This edition first published 2022 by John Wiley & Sons, Inc., 111 River Street, Hoboken, NJ 07030, USA and Scrivener Publishing LLC, 100 Cummings Center, Suite 541J, Beverly, MA 01915, USA

For more information about Scrivener publications please visit www.scrivenerpublishing.com.

Wiley Global Headquarters
111 River Street, Hoboken, NJ 07030, USA

For details of our global editorial offices, customer services, and more information about Wiley products visit us at www.wiley.com.

Library of Congress Cataloging-in-Publication Data

ISBN 9781119631644

Cover image: Pixabay.Com
Cover design by Kris Hackerott

Set in size of 11pt and Minion Pro by Manila Typesetting Company, Makati, Philippines

Printed in the USA

Dedication

This book series is dedicated to rotating machinery professionals around the globe who have devoted their careers to repairing, evaluating, and optimizing their equipment. It is through their diligence that critical machines are able to operate safely, efficiently, and reliably between scheduled outages.

Contents

Preface

Good reliability engineering is not the search for perfection—rather, it is the search for pragmatic solutions to business problems.

H. Paul Barringer, Reliability Consultant

Rotating machinery represents a broad category of equipment, which includes pumps, compressors, fans, gas turbines, electric motors, internal combustion engines, etc., that are critical to the efficient operation of process facilities around the world. These machines must be designed to move gases and liquids safely, reliably, and in an environmentally friendly manner. To fully understand rotating machinery, owners must be familiar with their associated technologies, such as machine design, lubrication, fluid dynamics, thermodynamics, rotordynamics, vibration analysis, condition monitoring, maintenance practices, reliability theory, etc.

The goal of the **"Advances in Rotating Machinery Series (3 volumes)"** book series is to provide industry practitioners a time-savings means of learning about the most up-to-date rotating machinery ideas and best practices. This three-book series will cover industry-relevant topics, such as design assessments, modeling, reliability improvements, maintenance methods and best practices, reliability audits, data collection, data analysis, condition monitoring, and more.

Volume 1 of the series focused on design and analysis. Volume 2 covers important machinery reliablity concepts and offers practical reliability improvement ideas. Best-in-class production facilities require exceptional machinery reliablity performance. In this volume, exceptional machinery reliability is defined as the ability of critical rotating machines to consistently perform as designed, without degradation or failure, until their next scheduled overhaul.

Specifically, Volume 2 contains the following content adressing machinery reliability:

1. General Reliablity Advice
2. Design Audits and Improvement Ideas
3. Maintenace Best Practices
4. Analyzing Failures

This volume provides readers with sound advice that can be used to assess and improve the reliablity of a broad range of process machiney, including process pumps and compressors, gas and steam turbines, and mechanical seals and their support systems. I hope readers will find this book to be a useful addition to their technical libraries.

Robert X. Perez, Editor
Fall 2021

Acknowledgements

I would like to thank all the contributors for their expert advice and their clear and insightful prose. Without them, this book series would not have been possible. I would also like to thank the publisher for believing in me and allowing me to develop this comprehensive book series. Finally, I would like to thank my wife for reviewing my drafts and for her encouragement.

Part I
GENERAL RELIABILITY ADVICE

1

Machinery Reliability Management in a Nutshell

By Robert X. Perez

How to Think Like a Machinery Reliability Professional

Figure 1.1 Managing the risk associated with a site's process machinery is a key role of a machinery reliability engineer.

An important role of a machinery reliability professional is to evaluate reliability performance of site machinery and provide clear feedback and recommendations to the owners to ensure process reliability goals are met (Figure 1.1). My aim here is to give the reader a sense of how machinery reliability professionals approach the assessment of site assets. By briefly covering the key machinery factors used in my evaluations, I hope to shine a light on the evaluation process. My process requires that each machine's criticality, history, and current site safeguards be carefully reviewed and evaluated. Only when this process has been completed can a machinery reliability engineer provide useful feedback that will hopefully have a positive impact on the site's overall reliability.

Robert X. Perez (ed.) Maintenance, Reliability and Troubleshooting in Rotating Machinery, (3–24) © 2022 Scrivener Publishing LLC

Criticality

The first step in a machinery assessment is to determine how the machine fits into the process, i.e., 1) What is the machine's function? 2) Is it unspared or spared? 3) If multiple machines are installed, then you must ask: Is there any standby capacity? A typical equipment description may look like this: A 2500 hp, multistage centrifugal hydrogen recycle compressor is unspared. If the compressor goes down unexpectedly, then there will be an immediate upset of the process, leading quickly to a unit shutdown. During this step, it is vital that you understand key machinery details, such as:

- Are you dealing with a dynamic, i.e., centrifugal, machine or a positive displacement machine?
- What are the types of bearings being used? Are the bearings fluid film bearings or rolling element bearings? How is lubrication delivered? Do the bearings require an external lubrication system?
- What are the types of seals being used?
- Is the train variable speed?
- How is the flow controlled?
- What are the current machinery alarms and trips and is condition monitoring equipment installed?

A thorough understanding of the machinery's construction and performance is required to conduct an in-depth reliability assessment.

Next, we need to ask, if the machine goes down or fails catastrophically, what are the consequences? A consequence is an undesirable effect related to a machine's failure or unavailability. Consequences fall into four basic categories: economic consequences related to process losses, environmental consequences, safety consequences, and economic consequences of machinery damage.

Table 1.1 lists typical economic consequences related to machinery operating in high-value processes. For example, a brief outage of one of several installed machines represents a low consequence event. However, if the same unit is down for several weeks, a significant economic loss could be experienced if production had to be curtailed for an extended time. The highest economic consequence is related to an extended outage of a single, spared machine. There is therefore a wide range of economic consequences related to process machinery. The goal of a machinery reliability engineer

Table 1.1 Economic consequence ratings of process losses related to machinery in high-value processes.

Equipment arrangement	**Brief outage of one machine (<24 hours)**	**Multiday outage of one machine (<1 week)**	**Extended outage of one machine (>>1 week)**
Single, unspared machine	Medium Consequence	**High** Consequence	**Extremely High** Consequence
Multiple partial capacity machines (no standby capacity)	Low Consequence	Medium Consequence	**High** Consequence
Multiple machines with one or more standby spares	Extremely Low Consequence	Low Consequence	Medium Consequence

is to dole out plant resources so that extremely high and high consequence events are rarely seen.

A common question is: What is the difference between consequence and risk? Machinery professionals generally define risk as *probability x consequence* or *rate of occurrence* (l) *x consequence*, i.e., Risk = P x C or λ x C. For example, if the consequence of an event is $1,000,000 and the event occurs once every 5 years, then the economic risk is $1,000,000 x 0.2 = $200,000 per year. When quantifying machinery risk, we normally use "rate of Occurrence (λ)" because machine failures can easily be tallied and converted into a failure rate. When it comes to rare failure events, we might have to review industry wide data or talk to equipment manufacturers to obtain realistic values. Using "rate of occurrence" risk equations for risk, we can see that to reduce a machine's risk level we must either, 1) reduce the frequency of occurrence for high consequence events, or 2) reduce the impact of high consequence events through the use of safeguards.

In addition to economic consequences, there are also environmental and safety consequences. Here are definitions for high- and medium-level events in these two categories:

Environmental Consequences

High — A major product release caused by machinery failure that impacts the community.
Medium — A significant product released caused by a machinery failure.

Safety Consequences

High — A major release of a toxic or flammable material caused by a machinery failure that impacts the community.
Medium — A significant release of a toxic or flammable material caused by a machinery failure could harm plant personnel.

Finally, we must consider economic consequences of machinery damage. In general, we know that larger process machines are more expensive to repair if they fail catastrophically than smaller process machines. For this reason, it is common to rank the economic consequences related to catastrophic failure in terms of a machine's rated horsepower. Perhaps, a machine rated below 200 hp can be considered to represent a low economic consequence if it fails catastrophically, a machine rated between 200 hp and 2000 hp can be considered to represent a medium economic consequence if it fails catastrophically, and a machine rated above 2000 hp can be considered to represent a high economic consequence if it fails catastrophically.

Consequences are additive, which means we can tally up all the various consequence to arrive at a cumulative consequence level. Let's continue with the example compressor at the start of this section and list all the consequences:

A 2500 hp, multistage centrifugal hydrogen recycle compressor is unspared. If the compressor goes down, then there is an immediate upset of the process leading quickly to a unit shut down. There are three categories of consequences related to a failure:

Economic consequences related to process losses:
Multiday Outage (<1 week) = High Consequence
Extended Outage (>>1 week) = Extremely High Consequence

Safety consequences:
A significant release of a toxic or flammable material caused by a machinery failure could harm plant personnel = Medium
Economic consequences of machinery damage

Table 1.2 Summary of potential failure consequences of interest.

Type of consequence	Low consequence	Medium consequence	High consequence	Extremely high consequence
Process Loss			X	X
Environmental		X		
Safety				
Machine Damage			X	

Since the compressor is rated at over 2000 hp, the economic consequences of a catastrophic failure are high.

We can now summarize our risk assessment, as seen in Table 1.2, and conclude the highest consequences are associated with process losses. Our summary only contains significant consequences of interest.

Machines associated with potential high consequence events are considered critical machines, while those associated with low consequence events are considered non-critical machines. In our example, the extreme consequences are related to extended outages and catastrophic machine damage. It makes sense to label this machine a high criticality machine in your listing of plant assets. When you are done with your site assessment, you should have a listing of all your machines along with their criticality ratings. If you force-rank the list, the machines at the top of the list should be your highest priority in terms of potential impact on the plant.

Equipment History

The second step of a rotating machinery reliability assessment is to review the machine's operating history. A historical review means to 1) review the history of site failures, and 2) discuss operating issues with plant personnel. The goal of this step is to determine the predominant failures modes associated with the machine in order to better understand which failure modes are key to controlling process reliability.

Let's go through a simple example. Let's say you are evaluating an electric motor driven, centrifugal pump that has two predominant failure modes: Mechanical seal failures and bearing failures. Through your research you discover that 70% of the failures are related to mechanical seals, 20% related to bearings, and the balance are related to other component failures.

The failure history also indicated the motor can run 5 years between repairs. It's clear that, if you want to improve the pump's reliability, most of your efforts should be applied around the mechanical seal.

A historical review is a great starting point for an assessment because it allows you to identify the vital issues and then focus on them. Going back to our centrifugal pump example, if you discovered that there had been 2 seal failures per year over the last 4 years, it would be abundantly clear where you needed to direct your analysis efforts. During the historical review, it is important to maintain a broad view of the equipment and system and not focus on a single issue in case there might be hidden issues affecting reliability. For example, a seal issue might actually be related to a control issue and may not be associated with a seal design deficiency. A poorly designed tank level control scheme may be causing pump cavitation, which in turn leads to a seal failure. This hidden control issue highlights why you need to also talk to plant personnel about what is going on around the equipment you're assessing. For example, an operator might tell you that there are times the pump seems to be pumping gravel. The key at this step of your analysis is to take a holistic approach to your evaluation.

To assign a risk level to a failure mode, you must also understand the type of failure distribution associated with each failure mode. Figure 1.2 shows that there are numerous failure modes seen around machinery. Machinery can experience bathtub, wear-out, fatigue, break-in period,

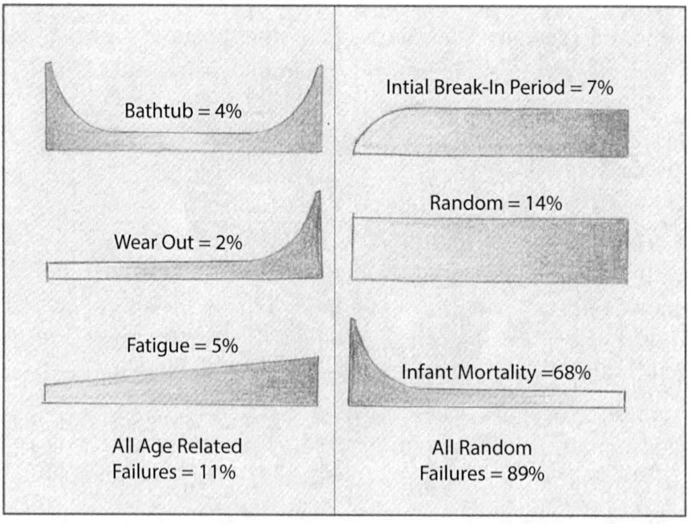

Figure 1.2 Machinery failure modes distributions.

random, and infant mortality type failure distributions. It is also possible to have a combination of these distributions.

When reviewing the failure distributions of a given machine, the question to ask is: What is the failure distribution associated with the failure mode of interest? Knowing the failure distribution is key to selecting the best maintenance mitigation strategy for each failure mode. The failure distribution of a failure mode tells you if the probability of failure is time related or not. If the failure mode is time related, then you might have a time-based activity aimed at servicing to replacing that component. If the failure mode is not time dependent, then a time dependent activity may not make sense. Your overall maintenance strategy is an accumulation of the best maintenance strategy for each historical failure mode.

Let's go back to our centrifugal pump example. Since the mechanical seal failure mode is the dominant failure mode, we must determine what its failure distribution looks like. Let's say that after studying the data, you conclude the failure distribution is random, i.e., not time dependent. (A technique called a Weibull analysis is commonly used to identify the failure distribution. Discussion of this analysis method is outside the scope of our discussion. To learn more about Weibull distributions and other reliability tools, visit http://www.barringer1.com/Contents.shtml.) It doesn't make sense to attempt to manage random failure distributions using time-based activities since they are not time related. One way to address machines with random failure distributions is to incorporate and install spare units to allow operators to swap pump in the event of an unannounced failure.

Here are some common strategies for the various failure distributions (see Figure 1.2):

1. Bathtub failure distributions: a) incorporate detailed start-up procedures to maximize the probability of a successful start-up, b) build in spare capacity to allow for random failures, c) incorporate a condition monitoring program that can detect early end-of-life failures.
2. Wear-out failure distributions: Use time-based inspections and replacements to minimize the risk of end-of-life failure.
3. Fatigue failure distributions: Use time-based replacements to minimize the risk of failure.
4. Initial break-in period failure distributions: Incorporate detailed break-in procedure to maximize the probability of a successful start-up and maximize the probability of successful start-ups.

5. Random failure distributions: a) build in spare capacity to allow for unannounced failures, b) incorporate a condition monitoring program that can detect early failures.
6. Infant mortality failure distributions: a) incorporate detailed start-up procedures to maximize the probability of a successful start-up, b) develop detailed repair procedures to minimize the probability of premature failures.

Detectable versus Undetectable Defects

A key concept in preventing potential machine failures is understanding whether an internal machinery defect is detectable or undetectable.

In Figure 1.3, the iceberg analogy is used to represent possible internal issues that might be found in rotating machinery. Notice that some failure modes represent detectable issues, such as high vibration or high temperature. I define detectable issues as those that can be sensed externally and easily monitored. However, there are also hidden issues, such as internal corrosion, erosion, stress corrosion cracking, etc., that can only be assessed through disassemble and inspection. During a machinery assessment, the analyst must tabulate both detectable and hidden failure modes and then

Figure 1.3 Machinery failure modes can be broken down into those that are readily detectable and those that are hidden from view. Note: failure modes related to high temperatures, i.e., creep, oxidation, etc., are related to gas turbines used as mechanical drives in compressor trains.

determine the best strategy to address them so that surprise failures do not occur.

Once all failure modes and their associated consequences are identified, then a comprehensive maintenance strategy can be developed. If the deterioration of a critical component can be measured and trended while a machine is in operation, then a condition monitoring strategy makes sense. However, if the deterioration of a critical competent cannot be assessed while a machine is in operation, then scheduled outages and inspection should be incorporated into the overall maintenance strategy.

A possible reliability strategy for a critical compressor might look like this:

1. Condition monitoring
    a. Continuous vibration monitoring with alarms and trips
    b. Continuous temperature monitoring with alarms and trips
    c. Continuous or periodic performance trending
    d. Continuous performance monitoring of compressors if the required instrumentation and analysis software are available

2. Periodic inspections
    a. Daily operator walkthroughs
    b. Weekly inspections by mechanics
    c. Quarterly performance evaluations by engineering if online performance capabilities are not in place

3. Scheduled Events
    a. Periodic borescope inspections of gas turbines
    b. Periodic water washing of gas turbines
    c. Cleaning (air-cooled electric motors, heat exchangers, fin fan coolers, etc.)
    d. Scheduled overhauls based on time in service or fired hours

There is no generic maintenance strategy for all the different types of rotating machinery. Each situation must be evaluated based on the historical failure mode, their consequences, and your maintenance budget. Some trial and error may be needed to identify the right mix of maintenance strategies.

Safeguards

A safeguard is any hardware and procedure designed to directly prevent an equipment failure or mitigate its impact. Typical safeguards include operating procedures, vibration monitoring systems, relief valves, overspeed trip systems, and surge control systems. One of the deliverables of a machinery reliability assessment is to determine if the current set of safeguards are adequate. If they are not considered adequate, then you must explain why they are not and provide detailed recommendations regarding any additions or modifications to the existing safeguards required to reduce risk to acceptable levels.

Compressor Operating Limits

An example of a safeguard is a procedure, alarms, or control system designed to prevent a harmful compressor operating condition. For example, variable speed centrifugal compressors have a maximum rotational speed limit. Above the maximum speed limit there is the potential to permanently damage the compressor rotor due to overstress. Every operating limit typically has a related adverse consequence, i.e., if a certain condition occurs, then something bad could happen. Examples of compressor operating conditions to avoid include:

1. Critical speeds
2. Minimum speed
3. Maximum speed
4. Surge flow
5. Stonewall flow (sonic flow conditions)
6. Maximum discharge temperature limit
7. Maximum vibration (% of bearing clearance) limit
8. Radial and thrust bearing temperature limits
9. Maximum horsepower limit

Compressor Flow Limits

Centrifugal compressors are designed to operate near their rated flow conditions. Operations away from the design point can result in unstable flow

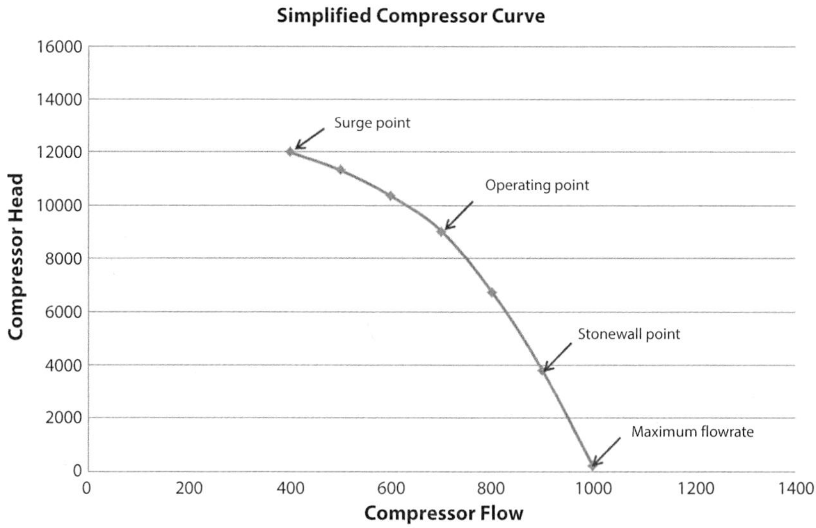

Figure 1.4 Centrifugal compressor operating limits. This compressor curve represents performance at a single speed.

phenomena. Figure 1.4 defines some of the operating conditions typically defined by centrifugal compressor manufacturers. Surge and stonewall conditions will be briefly touched on here.

Centrifugal Compressor Surge is defined as the operating point where the centrifugal compressor peak head capability and minimum flow limits are reached. Surge can be a detrimental condition in compression systems because it causes the compressor to vibrate violently due to rapid internal flow reversals, potentially damaging internal compressor parts. Surge control systems are specifically designed to prevent compressors from operating below the surge flow point or line. Figure 1.5 depicts a typical surge control system, which is composed of a flow measurement element, a controller with a surge control algorithm program, and a recycle valve. If, at any time, the inlet flow approaches the surge control line flow programmed into the surge controller, the recycle valve will open to increase the total compressor flow and prevent surge from occurring.

Centrifugal Compressor stonewall or choked is a flow condition that occurs when the sonic velocity of gas is reached at the exit of a compressor wheel. When stonewall flow is reached, the discharge pressure and flow drop precipitously. Flow through the compressor cannot be increased any further without an increase in the suction pressure. At the stonewall (or choke) flow the pressure versus volume curve becomes essentially vertical,

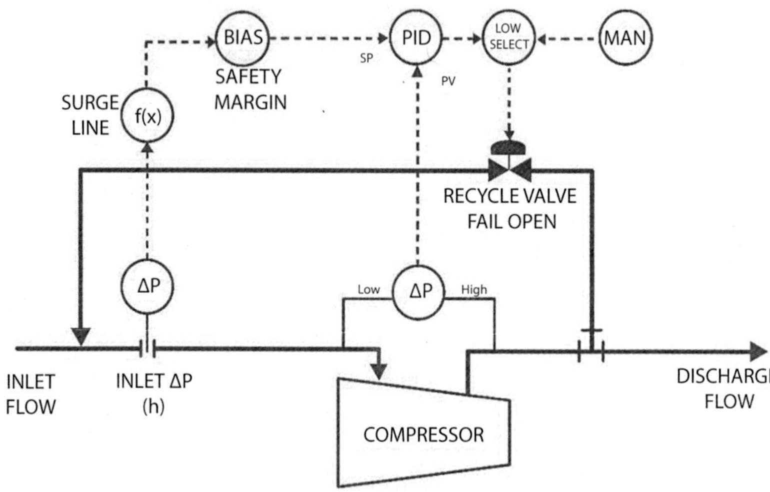

Figure 1.5 Typical centrifugal compressor surge control system.

and it is not possible to develop head or pressure at any greater flow. When the required operating flow exceeds the stonewall limit, the only remedy is to reconfigure the compressor with impellers (and matched stationary hardware) designed for larger flow rates.

Note: Axial compressors have been known to fail catastrophically at high flows due to blade vibration and high temperatures generated when operated beyond the stonewall line.

Critical Speeds

A critical speed is defined as any operating rotational speed that coincide with one of the rotor systems damped, natural frequencies. Operation at the critical speed for extended periods of time can result in internal damage due to excessive lateral vibration levels. It is important for operators to know their compressor critical speeds so they can be avoided. The original equipment manufacturer usually provides a list of critical speeds for users to avoid. Force response plots such as the one shown in Figure 1.6 can be used to locate critical speeds once the compressor has been installed in the field. The reader can see by studying the forced response plots in Figure 1.4 that rotor vibration increases significantly at about 1350 rpms and 2385 rpms due to the existence of rotor critical speeds.

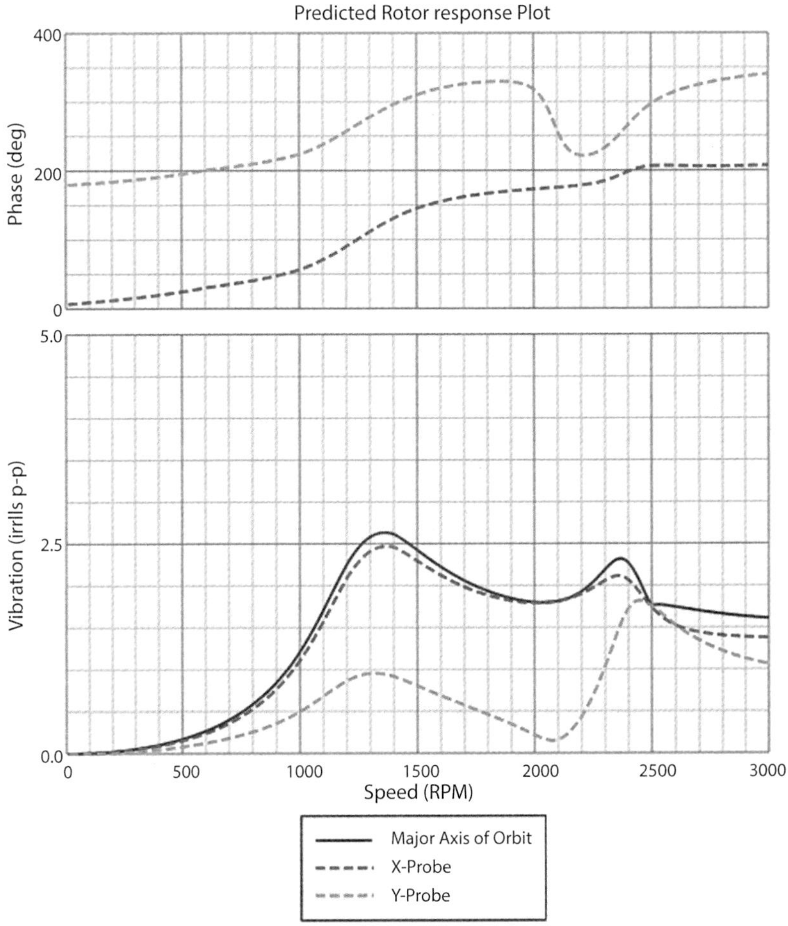

Figure 1.6 Here is a typical predicted forced response plot. The upper plot is the 1x phase versus the rotor speed and the lower plot is lateral vibration versus rotor speed. This analysis was run with mid-span imbalance. Notice: 1) there seems to be a rotor critical at about 1350 rpm where both the X and Y probes will see vibration peaks, which appears to be the rotor first critical, 2) there seems to be a second rotor critical at about 2375 rpm, 3) both critical speeds have major phase shifts associated with them.

Horsepower Limits

The horsepower load on a centrifugal compressor is governed by the flow, gas density, the rotor geometry, and the rotational speed. For a given set of impeller diameters and gas conditions, the fan affinity laws define how flow, pressure, and the required horsepower vary as the compressor's rotational speed changes as follows:

1. The flow rate varies directly with the speed
2. The differential pressure varies with the square of the speed
3. The power use varies with the cube of the speed

Small operating changes can have significant effects on how a compressor performs, which is why operators should carefully track the compressor's overall operating characteristics, including the horsepower load, and look for trends indicating process changes. An overload condition may be telling you something is wrong. Overloading may be manifested by high amp loads for compressor with electric motor drivers or a loss of speed for compressors with gas or steam turbine drivers. If amp loads are seen to be rising or dropping, try to determine the cause of the change. Rising amp loads may be caused by changing process conditions or machine degradation, depending on the shape of the compressor performance curves. Falling amp loads may be caused by a decrease in flow due to a flow restriction or a decrease in the suction pressure.

Temperatures

As most process gases get compressed, they heat up in a predictable way. For a given suction temperature, we should expect a predictable temperature rise due to compression at the discharge nozzle. However, deviations in predicted temperature values can occur, which is why operators need to keep an eye on compressor gas operating temperatures. Temperature levels provide vital clues to the condition of a compressor. High compressor discharge temperature or interstage temperatures can be a sign of compressor degradation or a higher than normal compression ratio, due to system fouling or a restriction.

> Takeaway: Gas temperature trending is a simple way of assessing the overall aerodynamic health of a centrifugal compressor.

Bearing temperatures can also provide vital clues to the condition of a centrifugal compressor. An upward trending bearing temperature is a sign that something is changing, such as the bearing's condition or the oil supply temperature. The most critical bearing temperatures are the thrust bearing pad temperatures. A failure of the thrust bearing pad will result in a catastrophic rotor failure as excessive axial movement occurs, which is why the axial position of the rotor should always be closely monitored and trended.

For an extra level of thrust bearing failure detection, a dual voting (2 out of 2, i.e., 2oo2) arrangement composed or two proximity probes are normally installed on critical compressors. (Note: The proximity probes must be installed axially so that they observe the thrust collar or other integral portion of the shaft surface. Probe targets, such as shaft sleeves, that are not integral to the shaft are not recommended.) Each machine train thrust monitor should be configured in a dual voting configuration, 2 out of 2 (2oo2), which is considered best practice by API 670 to ensure a high level of in machinery protection. In some highly critical applications, an OEM or end user may even request a triple modular redundant (TMR) configuration, 2 out of 3 voting (2oo3), to achieve the required reliability level.

A typical set of safeguards for a centrifugal compressor might include:

- A well-written set of operation and start-up procedures to prevent the compressor from operating at any unsafe flow or speeds
- Reliable flow and speed control
- A reliable surge control system
- A reliable vibration and thrust monitoring system
- Periodic lubrication monitoring
- Periodic field inspections of the compressor and driver
- Compressor performance monitoring

The purpose of safeguards is to prevent bad things from occurring. In cases where the consequences of an initial failure can escalate and grow in severity due to a progression of secondary damage, multiple safeguards can be used to mitigate the consequences of a failure by breaking the sequence of unwanted events. Figure 1.7 shows a hypothetical installation with a group of safeguards in place to prevent a failure or mitigate the effects of a failure. Without safeguards in place, a failure could progress from internal machinery damage to, in extreme cases, a catastrophic release with the potential to impact the plant or the surrounding community. In general, the greater the consequences of a failure, the more safeguards are required. We must always strive to achieve a balance between cost of safeguards and the risks involved.

Now let's use what we've learned and consider the risk matrix shown in Table 1.3. The far-left column shows event probability levels and across the top are event consequence levels. For example, the intersection of an "extremely high consequence event" and an event probability of "one event

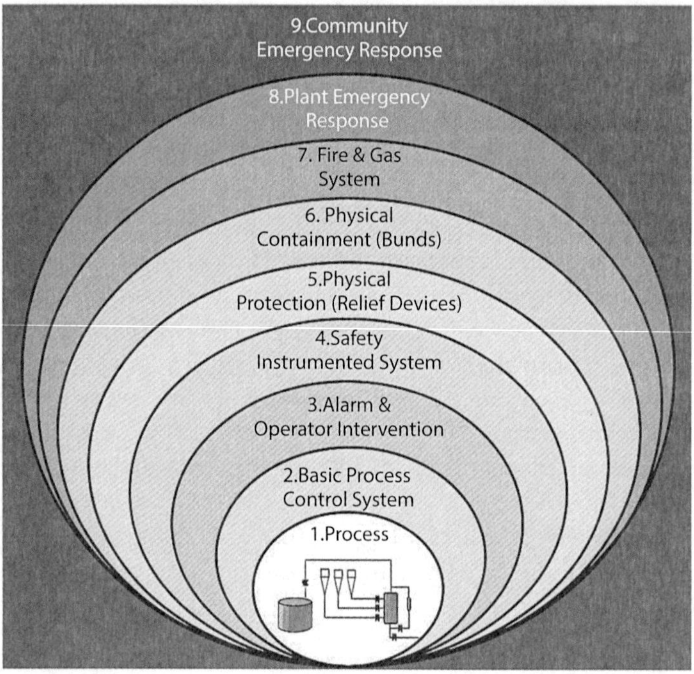

Figure 1.7 A set of safeguards are designed to prevent major events from occurring or to mitigate their effects.

Table 1.3 Here is a risk matrix for a hypothetical process machine. In this table, risk is defined as the frequency of an event occurring times the consequence level of the event (Risk = λ x C). Therefore, the cells in the upper-left corner of the matrix represent the highest levels of risk, and the cells in the lower-right corner represent the lowest levels of risk. Reliability professionals should strive to lower risk levels to acceptable levels using procedures, machine improvements or other safeguards.

Risk matrix	Severity				
Frequency of Occurrence (λ)	Extremely high consequence level	High consequence level	Medium consequence level	Low consequence level	Minimal consequence level
Several times a year	Not permissible	Extremely High	High	Medium	Low
One event per year	Extremely High	High	Medium	Low	
Event occurs every 1 to 10 years	High	Medium	Low		
Event occurs every 10 to 100 years	Medium	Low			
Event occurs every 100 to 1000 years	Low				

per year" represents an extremely high risk, as highlighted by the dashed oval. The idea of risk control is to design machinery systems so that all risks levels are at acceptable, low levels. If you find a particular risk level to be high during the course of your analysis, you can either reduce its probability of occurrence through machine redesign or by incorporating operating procedures or reduce the consequence level by adding additional protection, such as automatic shutoff valves, barriers, fire protection, etc. Moving the risk level to a cell that is either to the right or down from your starting point is desirable. An optimum outcome is moving your machine's risk in the general direction of the arrow in Table 1.3, which points down and to the right.

It is helpful to drill down to the failure mode level to understand the true nature of your risk. For example, if you have a mechanical seal that is failing more than one a year, causing a costly unit outage (high risk level), then you quickly realize you must improve the reliability of the seal. Other design efforts to reduce risk will likely be ineffective in significantly reducing the failure rate.

Layers of Machinery Protection

A set of safeguards should work together to provide multiple layers of protection (Figure 1.7) designed to eliminate failures and mitigate the effect of machinery failures. Here is a list of generic safeguards that can be used as a starting point for a review. As you go down the list, the consequences increase in severity. For example, items 1 through 4 represent normal operation, items 5 and 6 go into action if an upset has occurred and some type of automatic action is required for correction, and finally, items 7 through 9 are required when a product release has occurred.

You must decide if this list makes sense in your situation.

1. Detailed operating procedures are recommended in all cases. There should be specific operating procedure for start-ups, shutdowns, and normal operations.
2. A set of operating targets for flow, pressure, and temperature limits help keep your equipment away from unsafe operating conditions
3. Reliable process control systems
   a. Flow, pressure, and speed control
4. An alarm and operator Intervention plan
   a. Periodic walk-through inspections

   b. Machine alarms (temp, vibration, speed, etc.) to alert the operating of an unsafe condition

5. Safety Instrumented systems (SIS) for high consequence events, such as:
   a. Overspeed trip systems
   b. Temp and vibration monitoring system set to trip
   c. Automatic surge control

6. Physical Protection
   a. Relief valves for overpressure events
   b. Electrical breakers
   c. Mechanical overspeed trips

7. At this point of our hypothetical scenario, there has been a full loss of containment. The potential released volume and flammability of the liquid or gas released will dictate the severity of the event. Physical containments can be used for mitigation, such as:
   a. Isolation valves (with fusible links) & battery limit valves
   b. Dikes
   c. Ditches
   d. Barriers and walls

8. If the severity of a release is deemed great enough then a plant emergency response plan may be required.
9. If the severity of a release is deemed great enough then a community emergency response may be required.

When it comes to safeguards, the following adage fittingly applies: Plan for the worst but hope for the best.

Machinery Reliability Assessment Example

Background

Four 1200 rpm, 1500 hp, reciprocating, natural gas compressors (see Figure 1.8) are installed in parallel service. Each compressor is driven directly with a synchronous, electric motor. The gas processing facility needs all four compressors running to achieve design processing rates. If one of the

Figure 1.8 Reciprocating gas compressor cylinders.

compressors is down, the processing plant must cut back its throughput until it can be repaired and restarted. A process slowdown loses the plant about $50,000 per day when one compressor is unavailable.

There are several types of potential consequences associated with these four compressors. First, there is a significant economic penalty related to a process outage. The actual loss is equal to the length of the outrage times the length of the outage. A two-day outage will lose $100,000, while a 30-day outage represents a loss of $1,500,000. Second, there is a safety risk related to a major product release and subsequence fire scenario. Based on the existence of detailed emergency operating procedures and automatic isolation valving, the safety consequence level can be seen to be at medium. The third type of potential consequence is related to machinery damage. A catastrophic compressor failure could result in major damage and repairs costs. For example, a main bearing failure could potentially result in a crankshaft failure and even frame damage. If an undetected primary failure goes undetected, repair costs can escalate rapidly. In our example, since each compressor is rated at 1500 hp, we will rate the potential consequences of a major failure to be high.

After assessing all the potential consequences, we have rated the process consequence at a medium to high level, the potential safety consequences at a medium level, and the potential consequences of machine damage at a high level.

History

Our next step is to review the history of these compressors. During your review of the failure history, we discover that on average each compressor has experienced two valve replacements every year and therefore each one is down twice a year. A review of the valve failure data indicates that failures are time dependent, which suggests that valve monitoring may make sense as a way to mitigate the impact of a failure by allowing the process to better plan outages. There are no other significant compressor failure modes of interest. Furthermore, the synchronous electric motor drivers have demonstrated they can easily run between unit turnarounds without any need for maintenance. It is clear by reviewing the historical data that valve failures control the overall reliability of these machines. Interviews with operators indicate that there seems to be an acceptance that valve failures are the norm and that they have learned to live with the valve problems.

Safeguards

A review of existing safeguards indicates:

- There are adequate operating procedures in place
- Controls and safety systems are appropriate for the installation
- Field instrumentation are appropriate for these compressors. Note: there are thermocouples installed on all the valve covers to detect valve failures.
- The current compressor condition monitoring and lubrication programs seems to be active and effective
- There is proper management of all safety and reliability programs for these compressors

Conclusion

If we assume each valve failure leads to a two-day compressor outage, then the current consequence level is $50,000 per day times 2 days per outage, equals $100,000 per event, which we will consider a medium consequence level. Because there is a medium consequence level for a process outage and that compressors are failing at a rate of twice a year, we

Table 1.4 Risk matrix for machinery reliability assessment example.

Risk matrix	Severity				
Frequency of Occurrence (λ)	Extremely high consequence level	High consequence level	Medium consequence level	Low consequence level	Minimal consequence level
Several times a year	Not permissible	Extremely High	High	Medium	Low
One event per year	Extremely High	High	Medium	Low	
Event occurs every 1 to 10 years	High	Medium	Low		
Event occurs every 10 to 100 years	Medium	Low			
Event occurs every 100 to 1000 years	Low				

can conclude the current risk level is high (see cell with dashed oval in Table 1.4). It appears that the best way to reduce the risk level is to reduce the frequency of valve failures. Reducing the valve failure frequency from several times a year to once every 1 to 10 years moves the risk level from high to low.

Note: Installing an additional compressor to provide spare capacity is another way to reduce the overall risk level. However, a fifth compressor does not appear to be the most cost-effective solution to the current reliability problems. I will leave to the reader to analyze the pros and cons of installing a fifth compressor versus improving the reliability of the individual compressors.

In this example, the major recommendation is to review the present compressor valve designs to better understand why they are unreliable. First, the site needs to identify the root cause of the valve failures. Perhaps the failures are due to materials of construction, some type of gas contamination, or even the basic valve design. Every effort must be made to improve the service lives of these valves. Improving the reliability of these valves should be top priority for the site until their reliability of these four compressors are improved to acceptable levels.

Closing Remarks

I hope readers now have some understanding of how machinery professionals approach reliability reviews. We have discussed the three facets of machinery reliability assessments: A review of machinery criticality, equipment history, and safeguards. These three aspects of a machine are required to obtain a snapshot of the current state of reliability. After determining the criticality of a machine, we must evaluate the cause-and-effect relationships between adverse machine conditions and the various failure modes. We need to ask ourselves: Are these cause-and-effect relationships understood and repeatable? If they are, then we can begin to design proper safeguards to avoid future failures as a way of improving availability. For example, if we know that a low pump flow causes high vibration due to internal recirculation, which in turn causes a mechanical seal failure, then we may need to consider an automatic flow spillback valve to preclude low flow conditions. Keep in mind that every safeguard must be cost justified, i.e., the cost of each safeguard must return value in the form of reduced risk associated with the improvement.

Reliable operation requires: 1) proper physical safeguards, such as surge control, temperatures and vibration monitors, load monitors, speed sensors, etc., to prevent and detect unwanted operating conditions, 2) well-written operating procedures, and 3) judicious design improvements. We must be ever vigilant if we wish to keep our rotating machinery out of harm's way and operating reliably by avoiding undesirable operating conditions at all stages of a machine's lifetime. Before writing compressor operating and monitoring procedures, you need to clearly understand your machinery's operating limits. If in doubt, talk to the original equipment manufacturer about any concerns you may have. Always try to be conservative when setting operating limits. If you faithfully protect your process machinery with effective safeguards, you will be rewarded with many years of safe and reliable service.

2

Useful Analysis Tools for Tracking Machinery Reliability

By Robert X. Perez

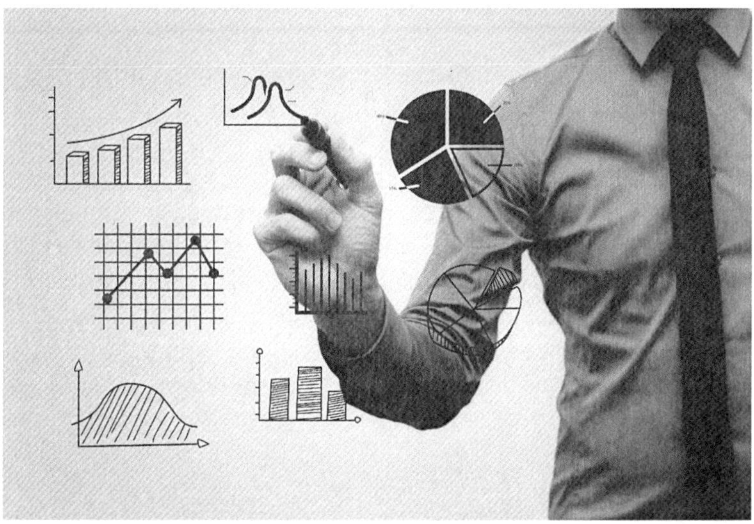

Figure 2.1 Machinery reliability metrics are essential to ensure that a site's reliability efforts are effective.

Reliable machinery is essential to realizing efficient and profitable plant operations. Most sites achieve world-class reliability through steady and continuous improvements in design, repair, and operating procedures. If we want to continue to improve the mechanical reliability of our machinery, we need to adopt and then maintain metrics and reporting tools that will help track and analyze how your equipment is doing with respect to

Robert X. Perez (ed.) Maintenance, Reliability and Troubleshooting in Rotating Machinery, (25–44) © 2022 Scrivener Publishing LLC

reliability (Figure 2.1). Carefully selected machinery reliability metrics and reports can:

1. Assist in identifying problem areas or applications.
2. Assist in identifying significant changes in reliability performance.
3. Document the impact of design improvements.
4. Document the impact of reliability programs.
5. Provide visual trend of machinery failures.
6. Allow us to compare the reliability performance of different sites and benchmark your overall performance with industry data.
7. Assist users in making economic evaluations of design modifications.
8. Assist in evaluating maintenance department productivity.
9. Assist in maintenance budgeting and planning.

When we talk about machinery reliability tracking, we typically break down the discussion into spared machinery and unspared, or critical machinery categories. Spared machines are less critical machine installations that have installed backup units (see Figure 2.2) to ensure the process is not interrupted in the event of one machine failure. On the other hand, critical machines are unspared machines that are essential to the operation of the process. If critical machines fail or shut down for an extended

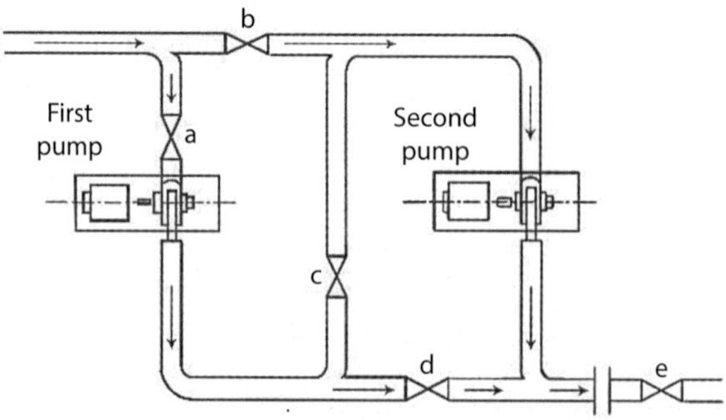

Figure 2.2 Main and spare pump arrangement is used to ensure reliability pumping service. If one pump fails, there is a standby pump ready for service.

period of time, the entire process has to shut down, resulting in significant economic losses.

Because these machinery categories expose the plant to different types of risks, their reliability performance is measured and reported differently:

1. Spared machines usually make up 90 to 95% of the plant population, which means they tend to consume most of the day-to-day maintenance resources. Therefore, the primary aim of tracking the reliability performance of spared machinery is to control maintenance resources, i.e., costs. When dealing with these large machine populations, we use reliability metrics such as MTBF, MTTR, and failure trends, bad actor lists, and planned maintenance percentage.
3. Critical machines make up a small percentage (less than 10%) of the population but tend to have a huge impact on the plant's operational reliability. Instead of maintenance costs, management is more concerned with production losses, environmental, releases, safety events, etc., which

Figure 2.3 The ultimate goal of machinery reliability metrics is to provide reliability professionals a way to reduce reliability data so that it can be presented in an easy-to-understand format.

can be an order of magnitude larger in consequence than repair costs. Therefore, when dealing with critical machines, we tend to use metrics such as availability, trends of process outages, cumulative downtimes (hrs.), production loss reports, Pareto downtime causes (machinery, exchangers, controls, etc.), Pareto of root cause, etc.

There is an endless number of machinery reliability metrics and reporting methods that can be maintained and reviewed (see Figure 2.3). However, in this brief survey, I will present tracking methods that I have used and believe have merit. The best metrics and reporting methods are those that are frequently used and have been proven to be useful in understanding the state of reliability at your site.

Commonly Used Metrics for Spared Machinery:

Mean Time to Repair (MTTR)

MTTR, also referred to as maintainability, measures the ability of a maintenance organization to restore equipment that has failed to a serviceable condition. Using MTTR, you can determine the average time it takes your maintenance staff to prepare, mobilize, and repair a machine that has failed and then get it back into service. MTTR is calculated as follows:

Total repair time/Number of repairs or replacements done

You can use this metric to find your site's current MTTR. If your current average repair time is unacceptable, you may need to look for ways to expedite the machine's restoration time. Reducing your MTTR can help decrease production losses resulting from maintenance downtime.

Mean Time Between Failure (MTBF)

MTBF forecasts the average time between one machine failure to the next under normal operating conditions. In other words, this metric can be used to predict the average life expectancy for a piece of equipment. To calculate MTBF, use the following formula:

$$MTBF = (M \times T)/R$$

Where M is the total equipment count, T is the reporting time, and R is the number of failures during the reporting time. For example, let's say we have 200 pumps in the population, and there are 20 failures in a 3-month reporting period. This would mean that the mean time between failures is 200 x 3/20 or 30 months between failures.

A higher MTBF is preferred. If one pump population (A) has a higher after MTBF value than another pump population (B), we can conclude, on average, the pumps in the A population will last longer in service than the pumps in the B population and are therefore more reliable than the pumps in the B population. Since the MTBF metric provides an idea as to how long the equipment will work without failure, it is a useful tool for forecasting repairs and replacement costs.

Note that sometimes MTBF and MTBR (Mean Time Between Repairs) metrics are used interchangeably. While they appear to be identical, they are not. The reason is that we may not know the exact time of failure, but we do usually know when a repair was done. Consider an example where one of a pair of spared pumps fails and is shut down and the spare pump is immediately started to keep the process running. It's possible that the failed pump may not be repaired for weeks, depending on the condition of its twin and the shop's workload. Therefore, we may not document the actual time of the failure, but we will know the time of repair. Make sure you know if the metric is based on failure data or repair data before using it to make decisions.

Additional Reliability Assessment Tools for Spared Machines

Useful reliability analysis tools take the available historical failure data and transforms them into either visual or concise tabular results. These visual displays can identify reliability problems that require our attention. Here are some of the types of reliability analysis tools we will cover and/or review:

- Pareto failure plots (for more information, see *Pareto Charts & 80-20 Rule* insert below).

- Bad actor forced rankings
- Reliability growth plots
- MTBF trends

Let's start by reviewing a simple tool that looks at failures on a sitewide basis. Table 2.1 contains a forced ranking of pump failures for various processing units across a site. By listing the mean time between repairs (MTBF) over the last 12 months, we can quickly identify the potential areas that may need addressing. In the case shown here, the Cat Cracking area seems to be the most problematic of all.

The pump failure data from Table 2.1 can also be converted into a Pareto chart to provide a summary of pump reliability at a glance (Figure 2.4). (A Pareto chart, seen in Figure 2.4, is a bar graph display of the frequency that events or measurements appear in a data group of interest.) In our Pareto chart example, pump failure frequencies over the last 12 months for various processing areas are plotted in order of decreasing failure frequency from left to right. Pareto charts are extremely useful for identifying issues that should be addressed first. The reader can quickly see that the Catalytic Cracking area had the most pump repairs over the last 12 months, and that

Table 2.1 A hypothetical table of pump failures across a processing facility.

	Number of pump trains	**Number of repairs last year**	**Total repair cost, $**	**MTBF (Months)**
Catalytic Cracker	50	34	272000	17.65
Coker Unit	42	21	168000	24
Crude Unit	40	15	120000	32
Alky Unit	35	15	120000	28
Fractionation Unit	40	15	120000	32
Sulfur Unit	25	15	120000	20
Utilities	42	12	96000	42
Hydrotreating	32	12	96000	32
North Terminal	30	8	64000	45
South Terminal	32	8	64000	48

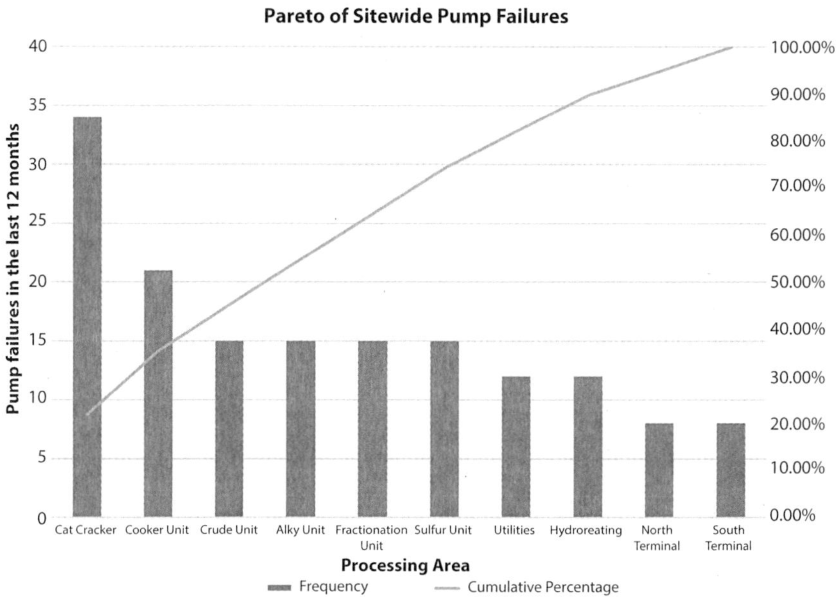

Figure 2.4 Pareto chart of total pump failures over the last 12 months for various processing units. The "cumulative percentage" line helps in determining how various groups add to the total failure population. For example, the Cat Cracker and Coker Unit failures represent about 35% of total plantwide pump failures.

Table 2.2 A forced ranking of pump failures in a hypothetical cat cracking unit.

Pump	Failures in last 12 months	Total repair costs for the last 12 months
31-P-09 A&B	5	$ 50,000
31-P-05 A&B	4	$ 40,000
31-P-04 A&B	3	$ 30,000
31-P-08 A&B	3	$ 30,000
31-P-17 A&B	3	$ 30,000
31-P-25 A&B	3	$ 30,000
31-P-02 A&B	2	$ 20,000

(Continued)

Table 2.2 A forced ranking of pump failures in a hypothetical cat cracking unit. (*Continued*)

Pump	Failures in last 12 months	Total repair costs for the last 12 months
31-P-06 A&B	2	$ 20,000
31-P-10 A&B	2	$ 20,000
31-P-11 A&B	2	$ 20,000
31-P-12 A&B	2	$ 20,000
31-P-14 A&B	2	$ 20,000
31-P-16 A&B	2	$ 20,000
31-P-18 A&B	2	$ 20,000
31-P-19 A&B	2	$ 20,000
31-P-22 A&B	2	$ 20,000
31-P-23 A&B	2	$ 20,000
31-P-01 A&B	1	$ 10,000
31-P-03 A&B	1	$ 10,000
31-P-07 A&B	1	$ 10,000
31-P-13 A&B	1	$ 10,000
31-P-20 A&B	1	$ 10,000
31-P-21 A&B	1	$ 10,000
31-P-24 A&B	1	$ 10,000
31-P-15 A&B	0	$ -

the South Terminal area had the fewest repairs over the same time period. The visual results from this Pareto chart suggest that more study of the catalytic cracker pump failures is warranted.

Now that we know most of the pump failures occurred in the Cat Cracking unit, we can narrow our focus to those pumps. Table 2.2 shows a forced ranking of the pumps with the most failures. In our hypothetical case, Pumps 31-P-09 A&B failed five times in the last 12 months. Assuming that each repair costs about $10,000, we now see that the worst actor cost us about $50,000 in the last 12 months.

You may choose to label the least reliable pumps at your site as "bad actors." Bad actors typically make up 7% to 10% of the pumps at your site that cost the most to maintain and cause you the most headaches. It makes sense to aggressively address bad actors first.

Pareto Charts & 80-20 Rule

The Pareto Chart is a very powerful data analysis tool that can be used to show the relative importance of problem areas and their root causes. They are composed of both bars and lines, where individual values are represented in descending order by bars, and the cumulative total of the sample is represented by the curved line. The 80/20 rule (also known as the Pareto principle or the law of the vital few and trivial many) states that, for many events, roughly 80% of the effects come from 20% of the causes. Joseph Juran, a well-regarded Quality Management consultant, suggested the principle and named it after the Italian economist, Vilfredo Pareto, who noted the 80/20 connection in 1896. Pareto showed that approximately 80% of the land in Italy was owned by 20% of the population. Pareto also observed that 20% of the peapods in his garden contained 80% of the peas. According to the Pareto Principle, in any group of things that contribute to a common effect, a relatively few contributors account for the majority of the effect.

Cumulative Failure Trends

Management is usually interested in knowing if their pump reliability is getting better or worse. A simple means of visualizing historical failure data is by contructing, then analyzing, a special trend called a reliability growth plot, which is a plot of cumulative failures versus time (see Figure 2.5). This type of graph is constructed by first creating a table of cumulative (total) failures in a population for consecuative time intervals, then plotting cumulative failures over the time period of interest. For example, let us say that in the first month 20 failures occur in a population, in the second month 25 failures occur, and in the third month 30 failures occur. This would mean the first three points in your reliability growth plot would be: Month 1, 20 failures; Month 2, 20+25=45 failures; Month 3, 20 + 25 + 30 = 75 failures, or (1,20), (2,45), and (3,75).

Reliability growth plots allow you to easily see tendencies in the failure data. Figure 2.5 shows three idealized reliability growth plots:

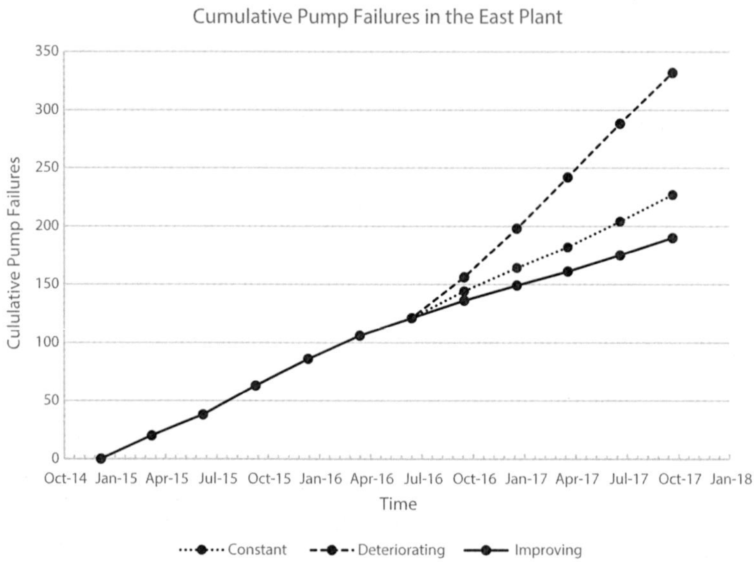

Figure 2.5 Reliability growth plot of pump failures in an operating area.

1. A trend where the slope of the cumulative failures is essentially straight, indicating a constant rate of failure (shown as "Constant" in Figure 2.5).
2. A trend where the slope of the cumulative failures versus time sharply increases in July of 2016, indicating a decreasing failure rate (shown as "Decreasing" in Figure 2.5).
3. A trend where the slope of the cumulative failures versus time decreases in July 2016, indicating an increasing failure rate (shown as "Improving" in Figure 2.5).

When readers study a reliability growth plot, they are able to discern if pump reliability is constant, deteriorating or improving. Other insights that a reliability growth plot provide are when a change in reliability occurred, and if the change in reliability was sudden or gradual. In this case, seen in Figure 2.5, we note the deteriorating case indicating that something changed after July 2016. We should look for changes in operating procedures, repair methods, processing rates, etc., to explain changes in pump reliability. Persistent changes in pump reliability may represent some sort of major change affecting your pump or pumps, while a data blip could simply be measurement error, or some sporadic factor, such as a plant upset.

The next reliability plot we will cover is the mean time between repairs (MTBF) trend plot (Figure 2.6). MTBF is a simple calculation that provides insight into the mechanical reliability of a single or a group of pumps, which is calculated as follow:

$$MTBR = \frac{Pumps \times Time}{Repair}$$

Let's go through an example of how we can evaluate the change in the MTBF of a large population. Let's assume you have 1,000 pumps in your facility and that in year #1 you repaired 375 pumps and that in year #2 you repaired 300 pumps.

The average MTBF for year #1 is:

$$MTBF = \frac{Pumps \times Time}{Repair} = \frac{1000 \times 12}{375} = 32 \ months\ between\ \ failures$$

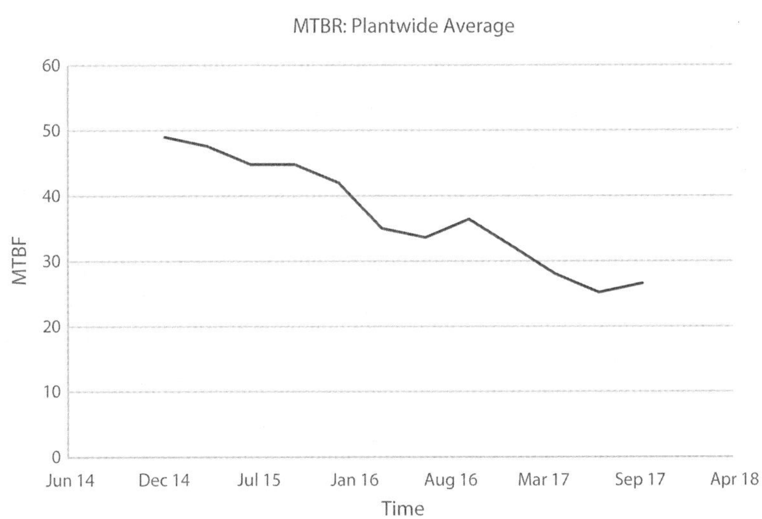

Figure 2.6 A hypothetical trend plot of the plantwide mean time between repairs (MTBF).

The average MTBF for year #2 is:

$$MTBF = \frac{Pumps \times Time}{Repair} = \frac{1000 \times 12}{300} = 40 \; months \; between$$

These data show a 25% improvement in the mean time between pump failures from year #1 to year #2.

Instead of continuously calculating the MTBF and reporting a single number to management, we can plot the moving average trend of a population's MTBF. This type of plot can be used to keep an eye on a single or large population of process machines. A moving average MTBF plot is constructed by calculating then plotting the calculated MTBF from the data from the previous month, quarter, or year. Plotting the MTBF data in this way tends to smooth out the data and simplifies the analysis. By inspection, we can see that the MTBF of a hypothetical pump population in Figure 2.6 is gradually deteriorating. Your next step might be to examine the individual trends of each of your process units to see if they are all deteriorating or if there are some poorly performing unit populations forcing the overall plantwide reliability down.

> MTBF: Readers should keep in mind that the MTBF metric is a global metric and only provides limited information about a given machine population. As you drill down to the equipment levels, more advanced analyses, such as Weibull analyses, may be warranted to examine the failure data and to better understand the nature of the failures.

I have briefly covered a few machinery reliability analysis tools that I have personally used to analyze spared machines during my career. The data used to create these metrics are readily available and can easily be organized and then analyzed using Excel or similar software applications. The concise visual results can then be quickly interpreted by your colleagues and management.

Metrics for Critical Machines

As mentioned earlier, critical machines tend to be regarded differently by management than spared machines because:

- Owners are usually more concerned about the availability of critical machines than their maintenance costs.

- There is a much smaller set of critical failure data than spared machine failure data.
- Production losses tend to dominate the economics of critical machines.

For these reasons, availability, trends of process outages, production loss trends, Pareto downtime causes (machinery, exchangers, controls, etc.), and root causes are often used to track the reliability of critical machinery.

Availability

Availability, also known as uptime, calculates what percentage of the time a piece of equipment was actually performed (or was able to perform if called upon by the site). It is an essential metric for measuring the overall effectiveness of an asset. If a machine's availability falls below a tolerable level, then the site must investigate the reason(s) why and develop a plan to improve it.

Normally, a tangible asset tends to lose its availability performance as it is used over time. Technicians can improve the asset's availability by undertaking essential maintenance tasks. However, when availability drops below a specific threshold, the asset's performance will not improve unless it is upgraded or replaced. There are several ways to calculate availability. The first is as follows:

$$\text{Availability} = \text{MTBF} \, / \, (\text{MTBF} + \text{MTTR})$$

An alternative means of determining availability is as follows:

$$\text{Availability} = \text{Uptime} \div (\text{Uptime} + \text{downtime})$$

Here is a simple example using the uptime/downtime method: A critical machine ran for 700 hours in a given month. During that month, the asset also had 12 hours of unplanned downtime because of a breakdown, and 8 hours of downtime for weekly PMs, which equals 20 hours of total downtime. Therefore 700 uptime hours + 20 downtime hours = 720 total hours. Using these numbers, we can determine the availability for the month is equal to 700/(700+20) = 97.22%.

Tracking availability can help identify opportunities for improvements by identifying problematic equipment. The typical availability benchmark is above 95% for most assets. However, it can differ depending on how necessary the equipment is to your operations.

Figure 2.7 Hypothetical machine history. Green (solid) arrows indicate machine is running and red (dashed) arrows indicate machine is down for maintenance.

Critical Machine Events

Without historical lifecycle data, we cannot make objective decisions about machines. To determine a critical machine's availability, we need to know how long it ran between outages and how long it took to make repairs. Therefore, we need a database that can capture life cycle events for all your critical machines.

Consider the life cycle of the hypothetical machine seen in Figure 2.7. At t_1 the machine is started for the first time. The machine runs reliably from time t_1 to t_2 and fails. The machine is repaired from time t_2 to t_3 and then restarted at time t_3. The machine runs from t_3 to t_4 and fails. The machine is repaired from time t_4 to t_5 and then restarted. The machine runs from t_5 to t_6 and fails at time t_6. The machine is repaired from time t_6 to t_7 and then restarted.

All of these start-up and shutdown times should be captured in machine records. In addition to these times, you also need to record:

- Repair or PM work scope performed during downtimes.
- Denote of the outage was planned or unplanned.
- Repair costs.
- Root causes of unplanned work.
- Losses incurred as a result of an unplanned outage. Losses can be related to production losses, demurrage, fines, etc.

If all this information is available, then the machine's availability (A_o), MTBF, MTTR can be determined and trended.

Some other tools used to track critical machines are:

Process Outage Trends

A production outage is when a processing unit is shut down due to a machinery, equipment, or processing issue. Unplanned production losses are especially costly. A great deal of effort is expended to ensure production losses are minimized. Production losses outages are typically reported in, 1) hours or days of downtime, 2) pound of production lost, 3) barrels of production lost, and 4) dollars of production lost, etc. If a production unit has no "make-up" capacity, then the production lost during an outage is lost forever.

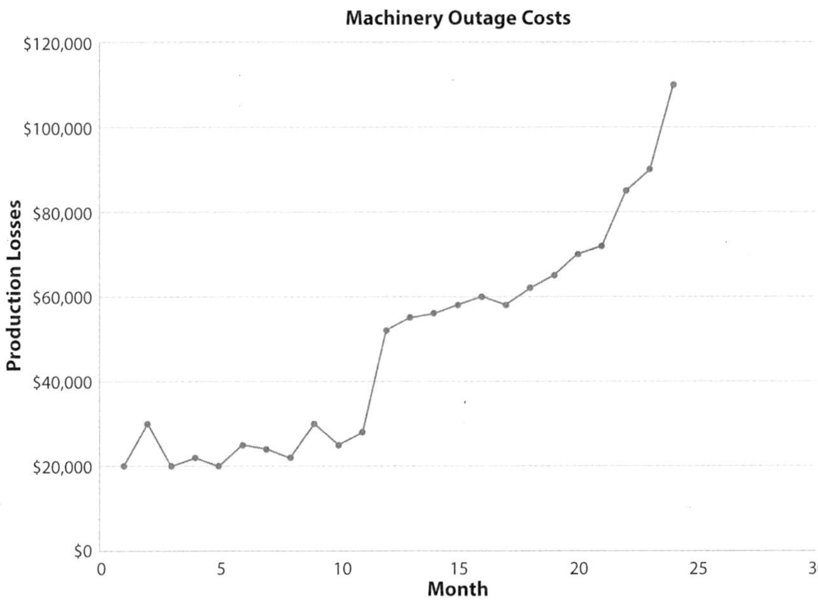

Figure 2.8 A trend of machinery outage cost will tell you the magnitude of the production losses over a period of time and if the losses are increasing, decreasing, or staying constant.

Here are a few types of plots that can be used to track process outages:

1. Production losses related to machinery outages (see Figure 2.8)
2. Breakdown of unit outages (see Figure 2.9)
2. Outage downtimes causes (machinery, exchangers, controls, etc.) (see Figure 2.10)
3. Root causes of outages (see Figure 2.11)

Process Outage Related to Machinery Outages

Pareto of Production Losses Across a Site

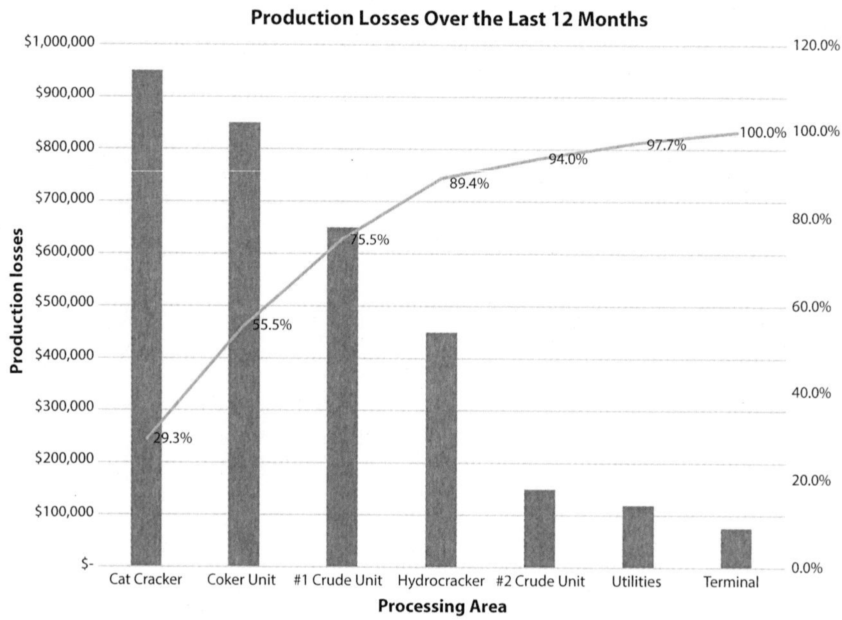

Figure 2.9 A Pareto of production losses across a site. This Pareto indicates that most (89.4%) of production losses are associated with four process units. The next step would be to drill down into this data to determine the root causes of these outages and address them.

Pareto downtimes causes (machinery, exchangers, controls, etc.)

A Pareto made up of RCFA causes can assist your organization in determining:

1. The best types of training that might benefit your organization.
2. How to best use your predictive maintenance resources.
3. Where design upgrades could be justified.

Note that to generate this type of Pareto chart your organization will need to perform RCFAs to determine the root cause of failures. RCFAs are essential for organizations that wish to have operational reliability.

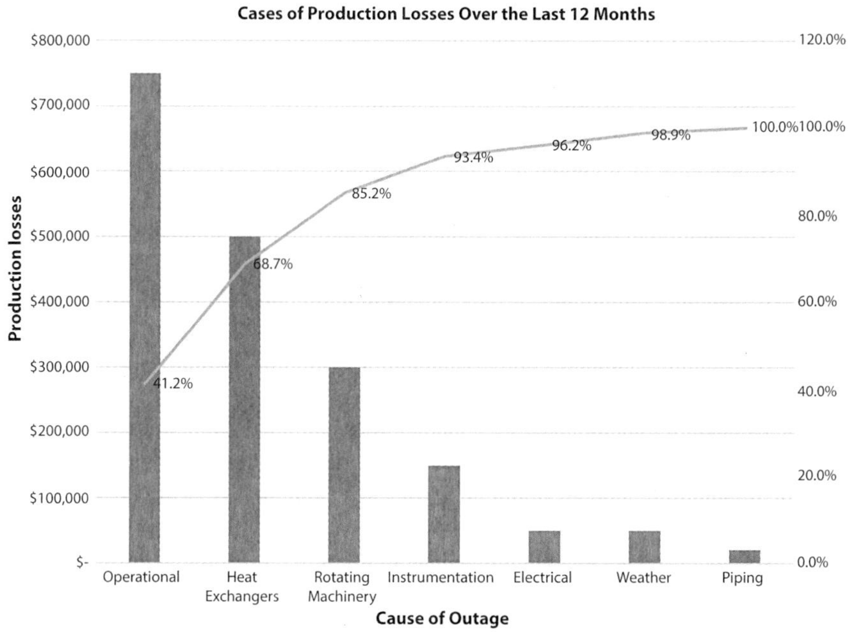

Figure 2.10 Pareto of causes of production losses over the last 12 months. This Pareto summarizes the reasons for unit outages. At a glance, management can see what is causing production losses. In this hypothetical example, we see that 68.7% of the process losses were caused by operational factors and heat exchanger issues. One conclusion you could draw from these results is that more operator training may be warranted to reduce process losses. Another recommendation that could stem from this Pareto is to review the present heat exchanger inspection program to see if improvements could be made.

Planned Maintenance Percentage (PMP)

Planned maintenance percentage (PMP), which is a reliability metric for both spared and critical machines, is calculated by dividing the total number of planned maintenance hours in a given period by the total number of hours spent on all maintenance in the same period. This number is multiplied by 100 to give the final percentage. The formula for the planned maintenance percentage is as follows:

$$\text{Planned Maintenance Percentage} = (\text{Planned maintenance time}/\text{Total maintenance time}) \times 100$$

For example, let's say the maintenance team of a chemical plant performs 500 hours of maintenance in a month. Of those 500 hours, 400 of them are planned maintenance hours. This makes our equation:

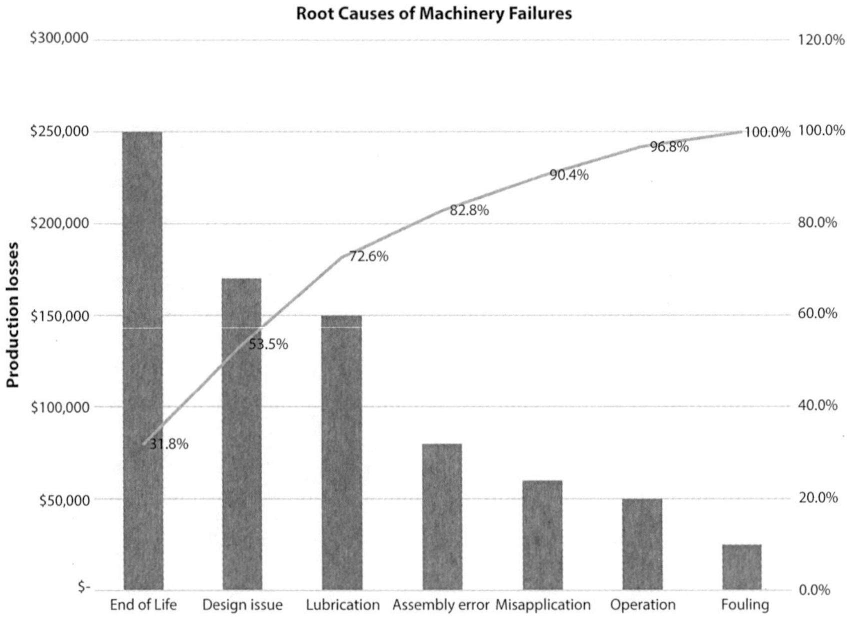

Figure 2.11 Pareto of machinery RCFA findings related to process outages. This chart points out the primary reasons for machinery-related process outages and can assist in optimizing your repair and PM strategies. In this hypothetical example, we see "End-of-Life" and "Design Issue" are the top reasons for machinery failure associated with unit outages. The "End-of-Life" results could suggest that you are either, 1) not identifying machines that are failing early enough to allow outages to be scheduled, or 2) you don't have the right inspection intervals identified for our critical equipment. "Design Issues" failures should be analyzed to see if economics warrant design upgrades.

$$(400 \text{ planned maintenance hours} / 500 \text{ total maintenance hours}) * 100 = 80\% \text{ PMP}$$

Ideally, at least 90% of planned maintenance time is necessary for optimal productivity. You can use the findings of Planned Maintenance Percentage to strengthen planned maintenance and reduce the likelihood of unexpected failures. There are numerous reasons to want to improve your facility's planned maintenance percentage. Some of these reasons might include wanting to:

- Reduce the amount of unplanned downtime at your site by optimizing equipment PMs.
- Lower maintenance costs by reducing overtime pay and expediting costs.

- Have greater control of facility budget due to planned costs as opposed to reactive costs.

Reliability Analysis Capabilities of your CMMS Software

Most maintenance organizations rely on some types of CMMS (Computerized maintenance management system) software to assist them in storing and analyzing the massive amounts normally generated by sites with sizable equipment populations. When selecting your CMMS software, it is important to clearly define the reliability metrics and reports you wish to create going forward. The software can only create the reports with data that is going to be available in the maintenance database. For example, if you expect to create Pareto graphs of equipment failure modes, the failure modes will need to be predetermined and the capability of entering failure modes needs to be available within your CMMS software. Since it will be costly to incorporate these database design changes after your software purchase, I recommend that these details are finalized early in your software selection and launch.

The right CMMS software allows you to determine:

- Are equipment failure rates getting better or worse?
- How well is your current preventive maintenance program working?
- Which machines are failing the most?
- What are the most common failure modes?

A major advantage of using an advance CMMS database is having the ability to drill down into the failure data. Take for example, the Pareto of the Machine Outage Causes. If you wanted to know which area experienced the most lubrication failures, you would drill down into the lubrication failure data to see if the failures were area specific or equally distributed across the site. If you found that there was a high frequency of failures in Unit A, then you might want to see of the failures were associated with a specific equipment type. However, to have this capability, you need to have the right database design and the data to draw from.

The right CMMS software capabilities will provide reliability metrics and reports on demand, which in turn will assist you in reducing the amount of work required to create useful reliability analysis reports. Having the right data, software, and predefined reports will greatly improve the likelihood of tracking your sitewide machinery reliability. Being able to detect major failure trends and specific machinery issues in a timely manner will allow you to judiciously direct the proper resources to your field problems.

3

Improving the Effectiveness of Plant Operators

By Julien LeBleu

Every plant has operators. They may be an untapped resource to improve reliability at a marginally low cost. Most operators believe it's their job to make on-spec product, do it safely, and prevent environmental excursions, and they are correct. However, management also expects them to detect problems early, gather operating data and troubleshoot field problems. Yet, many operators are not usually taught how to troubleshoot or given the tools that would allow them to perform these types of tasks effectively. Training on how to interpret pump curves or to properly inspect and troubleshoot rotating equipment is not always provided, but this view should be changed.

The savings on rotating equipment maintenance could be staggering if operators were trained to gather field data and troubleshoot proficiently,

Robert X. Perez (ed.) Maintenance, Reliability and Troubleshooting in Rotating Machinery, (45–60) © 2022 Scrivener Publishing LLC

so that problems could be identified early, thereby enabling them to avoid 10% or more of potential repairs or reduce chronic problems. Many times, the method of troubleshooting for maintenance and operators is to change parts until the issue is eliminated. It works, but it's a very expensive way to solve problems!

This chapter will primarily deal with pumps and seals. No matter how good your pump rebuilding capabilities are or how correctly the overhaul is done, through lack of knowledge the operator can cause equipment to be damaged or fail prematurely. Several days of training, ideally on a mixture of seasoned and new operators would result in the maximum benefit to the plant.

Items that will be Covered in this Chapter are:

- Look, listen and feel techniques [1].
- Applying look, listen, and feel techniques to field troubleshooting.
- Why the operator is important to troubleshoot.
- Operator tools.
- Operator understanding of the processing equipment, i.e., pumps, seals and sealing support systems with examples.
- A list of some equipment problems that overhauling a piece of equipment will not solve.
- Operator's involvement in capital projects.
- Writing quality work requests.

My goal is to stress that there are areas of training and utilization of operators that are not presently being taken advantage of. Everyone thinks operators are troubleshooting effectively and doing vigorous visual, audible, and tactile inspections of operating equipment in the plant. My experience and that of others whom I have talked with indicate that this is not true. Many operators do not understand lubrication, or how their equipment operates. They do not understand the relationships of pressure, flow, NPSH or horsepower, in pumps.

I will cover the areas where operators can be taught to identify problems early and how to relate it to equipment failures. Examples will be given to illustrate the techniques and benefits presented here.

Look, Listen and Feel

It is believed that operators already practice this inspection technique during walk around readings and inspections. It is true that some are very good at inspections, but the majority are not. They tend to walk through the plant and see the same thing over and over. Many times, unless sparks are flying, or smoke gets out of the machine operators may not notice key inspection clues. If things are changing very slowly, they may not notice it at all. (Boiled frog syndrome.)

Applying Look, Listen, and Feel Techniques to Troubleshooting

The following question seems never to be asked: "If you see something, what caused that and what would be the results if you were to do nothing?" For example, if an operator was walking by the tank, would he notice the black mark on the tank? (Figure 3.1a). If he did notice the mark, would he look for the cause, as seen in Figure 3.1b? It should be obvious that the hose is rubbing on the tank and eventually the hose will fail if nothing is done. The root cause is that the hose is too long and rubs the tank when pressurized. The hose needs to be shortened to prevent rubbing. If that is not possible because the equipment would have to be taken out of service, then perhaps the hose could be protected with an abradable wear cover to prevent failure. Remember, it is not always possible to solve the root cause of a problem, in which case acceptably treating the symptom is adequate.

Why the Operator's Input is Important to the Troubleshooting Process

Operators can add information to the work request that will make it easier to troubleshoot the problem or know what to work on. For example, if a specific problem happens repeatedly to a piece of equipment after a plant upset, temperature excursion, wash down, or some other recurring activity, the operator will likely be the only person to detect that, if he is attentive. Operators may gather valuable information that no one else has. For example, bearings may be installed in the pump correctly, but they are still failing after a wash down for a product change. The mechanic gets the

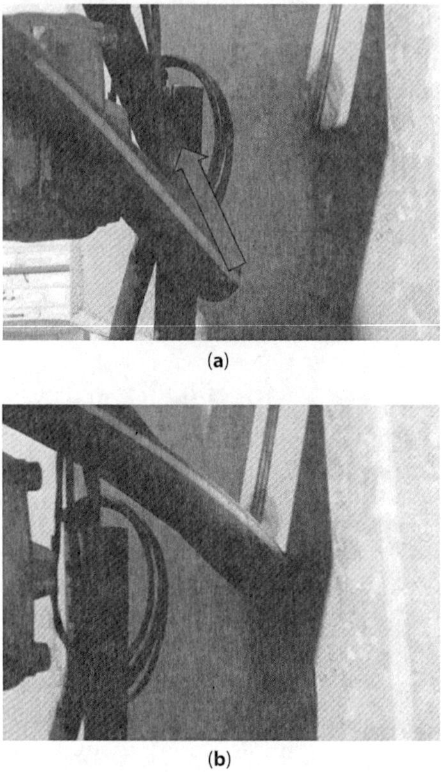

Figure 3.1 (a) Black mark on tank and (b) indication that the hose is rubbing on tank.

"bad press" even if he has done everything correctly because he was the one who installed the bearings and they failed. The operator can help by identifying this cause-and-effect relationship. An attentive operator will know what the noise, temperature, vibration, smell, etc., was yesterday, last week, and last year so changes can be detected and investigated.

Operator Tools

Operators should be given some tools to help quantify the inspection results they find. For example, touch will tell you something is hot, but that measurement is very subjective. How hot is it? Everyone may not feel the same "hot". Many companies give their operators an infrared temperature sensing device in the form of a small handheld tool as shown in Figure 3.2. They do not teach the operator its limitations, so many times the readings

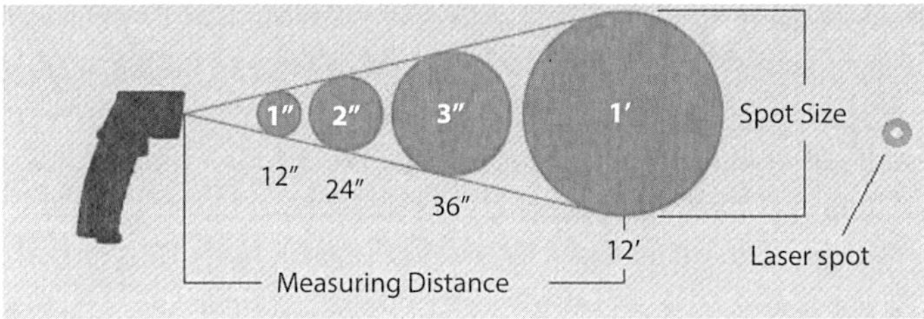

Figure 3.2 Note that the size of the temperature measuring area gets larger as the distance from the target increases. This characteristic of infrared thermometers can lead to measurement errors on localized hot spots.

are suspect. Many operators are not taught limitations of temperature for lube oil. The same situation exists with vibration pens or vibration limits. No matter how simple the tool is for operators, it must be explained and demonstrated and limits must be given for the tool, along with acceptable limits for the equipment.

One company started using a data gathering tool to collect data on equipment. One of the data measured was vibration. The device had alert settings. They were also instructed to call the vibration department for a second opinion, data gathering, and analysis when the alerts were reached. The vibration department was getting numerous calls at all hours because of the many alerts. The problem that accounted for the frequent calls was found to be inadequate training of the operators in the use of the vibration assessment tool. They were not told to place the probe on the equipment to be measured and allow the reading to settle down before pushing the button to take a reading. Pushing the measurement button too early resulted in a large abnormal vibration spike to be displayed and then recorded. After the training, the number of calls for additional vibration readings fell dramatically, and the readings were found to be more reliable. Then, whenever readings were found to be high there was a good chance a real problem was present.

These field measurement devices in conjunction with logic diagrams can reduce overtime substantially. With a temperature detecting device, vibration detecting device, and a strobe, quality decisions can be made, and quantitative data can be obtained.

Most operators do not want to make the decision to not overhaul something during the off hours, i.e., 6 p.m. to 6 a.m., weekends or holidays. The result is an "emergency" work order written to have a failed piece of equipment fixed immediately even if there is a spare and it is running well. Most operators are not aware that management is willing to accept some level of risk of the second piece of equipment failing during this time. The result is that operators make "very safe" decisions, so they are not second-guessed on Monday for what happened over the weekend. This is very expensive. One company instituted a plan where decision trees were generated regarding centrifugal pumps. A contact pyrometer, vibration detection device, and limits were given and explained. The operators were trained on the use of the tools and told to follow the decision tree or be able to explain why it was not followed. As a result of this change, making a good decision was easy. Two months after use was implemented the overtime dropped by 30%, which everyone agreed was very significant. Figure 3.5 is an example of a decision tree matrix that was used to achieve these results.

A strobe light is an excellent tool for doing field inspections while the equipment is running. Many problems such as broken fan blades on motors and finding impending coupling failures have been detected in this way. The source of seal leaks can be determined to be from the sleeve or the seal using this device.

Understanding the Equipment – Pumps, Seals and Sealing Support Systems

Operators need to understand the centrifugal and positive displacement pump curves. Without the relationships demonstrated in the curves, it is not possible to adequately troubleshoot the equipment. For example, if the discharge pressure in a centrifugal pump is increasing, the flow will likely

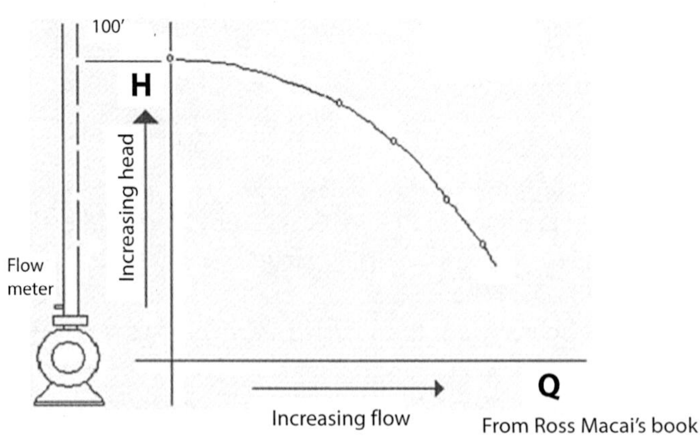

Figure 3.3 Basic centrifugal pump performance curve [2].

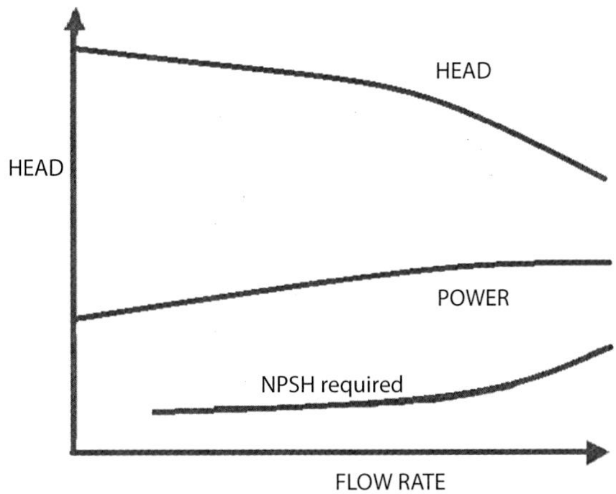

Figure 3.4 Typical centrifugal pump head versus flow, power versus flow, and NPSH versus flow curves.

be decreasing. That does not mean the pump must be overhauled, but it does mean there is likely a blockage downstream, as one example.

Figure 3.3 is an example of what can be taught to operators to explain the centrifugal pump curve.

In a hypothetical world, if the pump puts out 100 feet of head and a hole was drilled at an elevation of 100 feet above the pump, there would be liquid but no flow. If a hole were drilled into the pipe pump 10 feet below the first hole, there would be some flow, but 10 feet of pressure head would be lost. If holes were drilled every 10 feet down the pipe, more flow would result, but there would be a gradual loss of head as you move down the vertical pipe. This shows the relationship between head and flow in a centrifugal pump in the simplest terms. Figure 3.4 shows all of the interacting parts of the centrifugal pump curve.

Centrifugal Pump Relationships to Remember

Relationship of viscosity and head

- Viscosity up = head down

Relationship of viscosity and horsepower

- Viscosity up = horsepower required up

Relationship of flow and head

- Flow up = head down
- Head up = flow down

Relationship of flow and horsepower

- Flow up = HP up
- Flow down = HP down

Relationship of flow and NPSHR

- Flow up = NPSHR up
- Flow down = NPSHR down

Positive Displacement Pump Relationships to Remember

Viscosity to horsepower

- Viscosity up = HP required up
- Viscosity down = HP required down

Flow and pump speed

- Speed up = flow up
- Speed down = flow down

Speed and horsepower

- Speed up = HP up
- Speed down = HP down

Speed and NPSHR

- Speed up = NPSHR up
- Speed down = NPSHR down

Pressure and horsepower

- Pressure up = HP up
- Pressure down = HP down

Here is a list of some of the items that can affect a pump's performance that overhauling a pump will not fix.

- Plugged suction strainers can lead to pump cavitation, which sounds like gravel/marbles, discharge gauges fluctuations, and a lower than normal power draw.
- Closed valves (very rare, but gates have been known to come off their stems).
- Sediment in bottom of vessel blocking suction line – same as plugged suction.
- Vortex breaker failed, covering part of the suction line in the vessel.
- Not enough suction pressure – tank level low, unusually high product temperature causes pump to cavitate.
- Product too cold or too viscous (will not flow well enough to provide NPSHr/suction pressure – pump cavitates).
- Product too hot (product flashes at low pressure in the pump, causing pump cavitation or seal to run hot and fail).
- Air leaking into suction either at either the a) inlet flange, b) bleeder valve on suction not closed completely, c) plug not installed in a valve on the suction line, d) crack somewhere on suction side (vacuum service) can all cause the pump to cavitate or not pick up suction.
- A leaking check valve on standby pump can result in not enough flow/pressure on the main pump (spare pump may be turning backwards)
- Pump turning backward (suspect any time the electrical system has been worked in that circuit/area) results in approximately 40 – 50% of the normal flow
- Wrong pump speed, especially on steam turbine/air driven, variable speed equipment, or when pulleys are involved—the pump does not generate the rated flow or pressure.
- Blockage downstream of the pump (system pressure too high on discharge side of pump—the pump has high discharge pressure and low flow and may be the same discharge pressure as if the discharge valve is closed. When you start the stand-by pump there is no change in conditions. Low power draw in this case. If discharge strainers are used, they may be plugged
- Too much flow being required from the pump (rates have been increased from the original design or some type of

bypass flow has been instituted)—the pump is not capable of putting out the desired flow resulting in the discharge pressure being very low, and amps or power draw will be very high.

- Too little flow being required from the pump (causing high recirculation – pump too large – the pump has high pitch, tone type noise, pump casing may become very hot)
- Piping and or Structural Resonance (High Vibration)
- Product vaporization point is too close to suction pressure—this condition will cause cavitation, low, or no flow and may be most pronounced during an upset of the plant or perhaps a hot restart of the plant.

All of these should be checked before condemning and overhauling equipment.

Mechanical Seals

Operators need to be taught that seals are like bearings in their car. If they do not get cooling and lubrication, they will fail quickly. The support system, such as barrier fluid pots, piping to the seal and cooling water in them are required to keep the seal functioning. The support system is like the radiator and lubricating oil in the vehicle. If either are stopped for long a failure will result. The operator needs to be taught how to detect problems while equipment is running, such as looking at the differential temperature between the lines going in and out of the seal chamber of a double mechanical seal. If there is no temperature differential, then there is no heat exchange, which ultimately leads to a seal failure.

Most operators do not know that ALL seals leak. When a single mechanical seal leaks, the atmospheric leakage will always consist of the process liquid. However, in double seals, which have two sets of sealing faces, there are two possible leakage paths, depending on the seal configuration chosen by the seal designer: 1) a pressurized double seal can leak barrier fluid into the process and into the atmosphere; 2) an unpressurized double seal can leak process liquid past the inner seal, which contaminates the barrier fluid, and leak barrier fluid into the atmosphere. The takeaway here is that operators need to learn something about their mechanical seals so that they will be able to identify when they are leaking.

Most operators do not know how expensive seals are and would pay more attention to them if they did. Here is an example of how costly a lack of understanding and knowledge of mechanical seals can be:

> *An operator was working in the tank farm area of a plant. He needed to transfer the contents of one of the tanks to another tank for blending. The road that led to the tank the operator must supply to the suction of the transfer pump was about a quarter mile from the pump. On the road to the tank was the pump. The operator stopped the truck, got out and started the pump, then got back into the truck and drove to the tank supplying suction to the pump and opened the supply valve. This pump had many seal failures because of something the operator did not understand. If only the seal and its installation were investigated, the lack of operator knowledge would never have been addressed or the problem solved.*

Capital Projects

Invite operators to provide input on how new equipment can be efficiently decontaminated and vented, and how new valves can safely be accessed. This will also get agreement from the operators because their concerns are being addressed. It is a good practice to include a maintenance representative in design reviews as well.

Writing Quality Work Request

Most operators do not understand that work requests are one-way communications and that the closed work order will go to the history of that piece of equipment. If all the work request says is "change pump, replace pump, fix pump" there is little value added to the equipment history. It is only known that the equipment was worked on. If the work request clearly describes the symptom or the reason the equipment was removed it would be more beneficial to anyone investigating the equipment history. For example: If the work request read "too low a flow, broken coupling, vibrating, tripping offline," etc., this would help identify the recurring problem. It would make tracking the real reason for the failures easier to identify. This might also increase the cooperation between operators and craftsmen. Mechanics do not appreciate operators telling them what to do with their machine. However, mechanics do want the operators to define why they think there is a problem with the piece of equipment so the root cause can be determined and eliminated, which is better for everyone.

Procedures (Procedures and Decision Trees)

The use of procedures for startup and shutdown of equipment is essential. These can be helpful on common equipment just as they are for large, specialized equipment. Some companies use decision trees for common problems on equipment to make it easy for the operator to make the correct decision (see Figure 3.5). If they follow the procedure or decision tree they will not be second-guessed later. If they choose not to follow the decision, then there should be some sort of consequence for that choice. I will discuss that more in the "Discipline" section. These can be beneficial for mechanics and engineers as well.

Many operators do not know acceptable limits on oil temperature, bearing temperatures, or vibration levels. It is impossible for them to make informed decisions without having been taught what these operating limits are. Many operators do not fully understand the differences between these measurements or their ramifications. For example, the severity of the damage that can be done by excessive thrust is far worse that that done by excessive vibration. Many operators treat them as the same, but they are not.

Must Give Operators Feedback

The single largest complaint I hear is, "I am asked to give input but none of my suggestions go anywhere. I never hear what happened to my idea." Oftentimes the idea is used, but it is part of a capital project, or it may not be used for a specific reason. The operator's problem is that the reason some suggestions are not implemented are never explained to him, so from his perspective, he is asked for something, he complies, but never hears any more about it. This is not conducive to getting any other information from an operator.

One meeting concerning operators involved them making numerous work "safety" requests because they knew these would be investigated and followed up on. The operator wanted at least feedback as to why his observations were not being followed up on. His observation dealt with a steam trap leaking and blowing steam, water, and debris everywhere. It also caused the safety glasses to fog, impairing vision and making the area

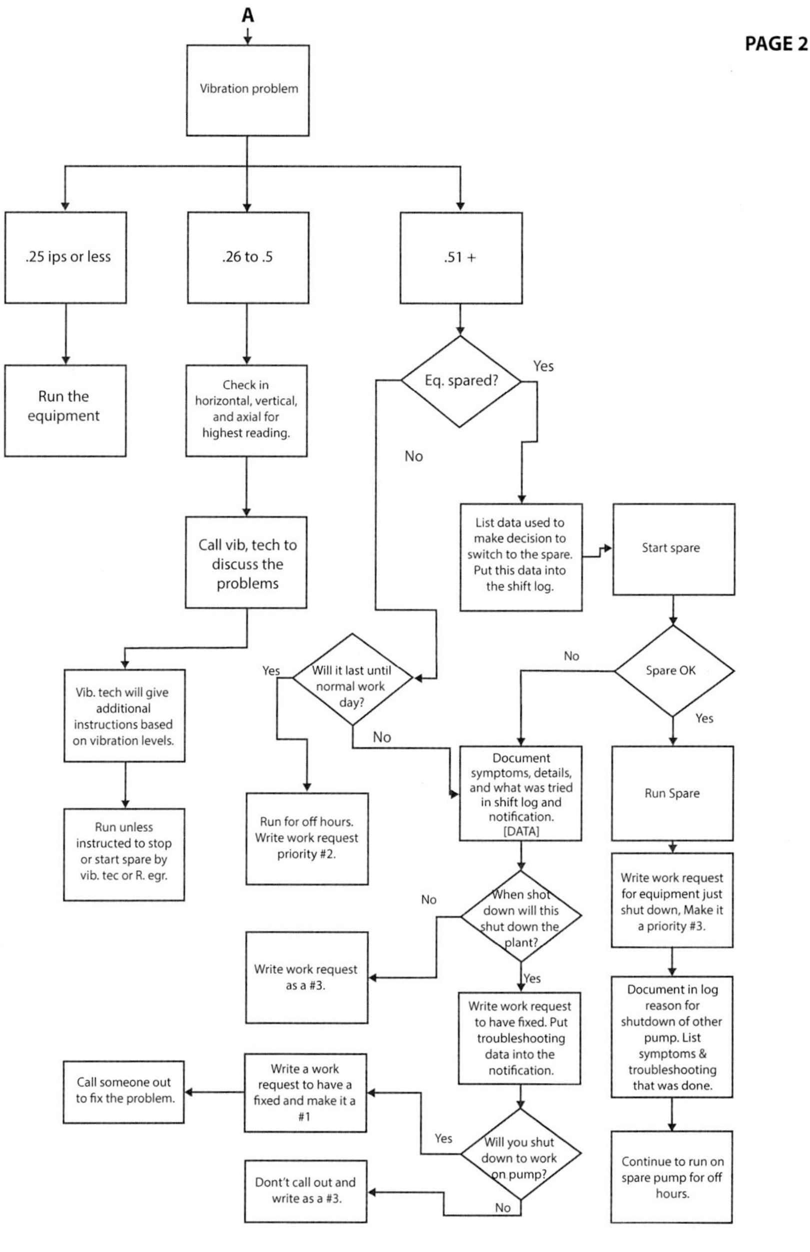

Figure 3.5 A portion of the decision tree that operators can use to make better decisions in the field during weekends and holidays. The full decision tree includes five subbranches: suspected vibration issues (seen here); suspected seal issues; oil leak issues; suspected bearing temperature issues; and suspected performance issues. Depending on the nature of the problem, the operator refers to the appropriate branch for instructions.

slippery. The operator said he had mentioned the problem months ago, but nothing was done. The maintenance manager in fact had a good reason for not acting: there were limited resources and either customers would be deprived of their order (force majeure), or the steam leak would be fixed. But the operator was not informed of this. When it was finally explained to him, the operator was still not happy, but he did understand the situation. With further talks, the operator offered to "rope off the area and also direct the steam spray in a safer direction."

Must be Required to Use their Training

Many companies have in-house and external training for the operators and other crafts. Someone from management should attend the training to know what is being delivered. They can then help ensure that the training is being utilized. Most consultants and trainers are happy to take your money, but for you to get the benefit of the training they offer, the concepts and methodologies must be used. The example used about centrifugal pump pressure rising and thus, the flow falling, should prompt the operator to look for a blockage downstream, not have the pump overhauled. If instead, the operator is allowed to do what he usually does, which is to write a work request and make the problem someone else's, it is possible that the pump may get overhauled only to find out there was a blockage downstream. Identifying the problem and working on the root cause can be helpful in increasing reliability.

Discipline

I am not in favor of firings or time off as a form of discipline except in the most serious cases. You lose a trained employee and perhaps someone that "won't make that mistake again." You may be giving a person time off to go hunting or fishing, which may not seem like discipline to him. I have found that if someone is made to give a formal presentation on how the decision was made, it usually results in the ultimate desired effect of better decisions. Additionally, when word spreads in the plant that faulty decisions will have to be formally explained, those who are responsible for making decisions will make better ones. Most people would not like to be required to explain, in a formal presentation, why or how they made a particular decision. Decision making will improve quickly.

Conclusion

As I have pointed out in this chapter, there are many opportunities to make the operator an integral part of the plant's reliability program by involving him in the problem identification and troubleshooting processes. Many people do not take advantage of what their operators have to offer and as a result their equipment reliability suffers. Most operators want to do a good job and take pride in their work but may not have been given the tools or training necessary to do their best job.

Appendix A References

1. Le Bleu, Julien, Perez, Robert, 2014, *Operator's Guide to Rotating Equipment.*
2. Ross C. Mackay, 2004, *The Practical Pumping Handbook.*

4

Spare Parts Strategies for Optimizing Rotating Machinery Availability

By Robert X. Perez

Figure 4.1 Bearings, seals, impellers are a few examples of rotating machinery subcomponents that need to be spared so that they are readily available for normal repairs.

Two key questions to ask after selecting a critical rotating machinery for your site are:

- Which spare parts should I stock (see Figure 4.1)? Buying too many spare parts wastes money. Buying the wrong mix of spare parts or too few spare parts could lead to costly extended outage times.
- What spare part management strategy makes the most sense? Not properly managing your spare parts can lead to overstocking or a loss of valuable parts. Today, most

Robert X. Perez (ed.) Maintenance, Reliability and Troubleshooting in Rotating Machinery, (61–74) © 2022 Scrivener Publishing LLC

warehouses seek to minimize the value of spares to reduce property taxes.

When trying to understand your spare parts requirements, you must first understand the economic consequences associated with not having the right parts in the warehouse when they are needed. Let's briefly cover some of the most common economic consequences related to a shortage of spare parts inventory:

1) The costliest process-related consequences are associated with an extended outage of an unspared machine required to keep a process up and running, such as an unspared compressor or charge pump. The penalty for downtime is determined by the daily value of the production plus any additional outage penalties, such as demurrage cost, delivery delay penalties, etc. If, for example, a spare part or assembly has a 60-day delivery, then the economic consequence of an extended outage is 60 times the daily production value plus any additional penalties. If the part or assemble delivery can be expedited, then the expediting cost must be added in the total losses. In the worst cases, an extended outage could even lead to the loss of an order or even a customer.
2) Next come less costly process-related consequences associated with an extended outage of one of multiple units operating in parallel. In these services, if one of the units is down, then production is partially curtailed. If the unit or plant if able to stay online at the lower flowrate, then this type of event would represent a medium level of risk. Let's take the example where there are three hydrogen compressors in a process and all three compressors are required to operate to run at full rates. If one of the compressors is down, then the process unit will only be able to operate at two-thirds of its full capacity. If the unit can operate at this reduced rate, then the economic penalty will be one-third of the daily production value for the entire time required to get the spare and repair the machine. However, since "slowdown" events typically do not incur product losses, energy costs, overtime costs related to a full shutdown and startup events, they are much less costly than process outages related to unspared machines. One approach to spares for multiple machines is to store common spare parts for the group. For example, you

may elect to store a single crankshaft for multiple identical reciprocating compressors at a site.

3) Then, there are less costly process-related consequences associated with main and spare machinery applications. The idea of using this type of arrangement is that if the main process machine goes down then an installed idle machine will be immediately available for service. The risk involved in this situation is that if the main unit is being repaired, then the installed spare unit is essentially running unspared for the time the main unit is being repaired. If the machines are reliable, i.e., have mean time between repairs (MTBR) > 5 years, the risk of the running the spare for an extended period is minimal. However, if the pumps have been proven to be unreliable, i.e., MTBR<<5 years, then the risk of losing both pumps is probable and quantifiable. In situations where the main and installed spare machine's reliability is deemed questionable, the warehouse may opt to keep higher levels of frequently used parts in stores to reduce risk levels (see Warehousing Example #3 below).
4) Finally, there are machinery that have very low or no economic consequences when they are down. This category, which is composed of smaller or non-process-related machines, typically justify very little or no spare parts inventory.

Spare parts management is a balancing act between minimizing the cost of acquiring and maintaining warehouse stores and controlling outage risks by having the right critical spare parts available when needed. One way to define risk is by multiplying probability of an event occurring times the economic value of consequence of the event (Risk=Probability x Consequence). We are betting that having the right parts on hand will significantly reduce the probability of an extended repair outage to an acceptable level. The probability of failure approach to risk is used in applications where failures are rare events or have never occurred.

Risk can also be defined as the frequency that an event occurs times the economic value of the consequence (Risk=Frequency x Consequence). For example, if a $100,000 outage occurs twice a year then the risk of this recurring event is $200,000 per year. The event frequency approach to risk is used in applications where outages occur frequently and are well defined. Both ways to quantify risk can be used to justify the optimum spare part stocking levels.

Here is a series of questions that will help you decide stocking levels for plant machinery:

1) Is the machine criticality low or high? Spared rotating machinery tend to be less critical and unspared rotating machinery tend to be highly critical. If you are not sure, ask the question: If this machine is down what are the consequences? If the consequences are costly process losses, then the machine is critical.
2) Is the spare parts usage rate predictable?
    a. Historical usage provides a clue to future usage, i.e., If a part fails consistently every 5 years and it takes 2 months to restock the part, then a low stocking level is justified.
    b. If a machine is taken out of service every 3 years for inspection, then you have a clue about when replacement parts will be needed.
    c. If spare parts usage is not understood, perhaps a statistical analysis of the data is required to understand when parts will be needed.
3) Is the machine repairable?
    a. Normally smaller machines, such as small motors or pumps, are not repairable (see Figure 4.2). If the item is not repairable you can either stock a complete spare unit in store or order it from the supplier when required. Obviously, if a spare unit is not readily available, then it makes sense to stock a complete unit in stores.
    b. In contrast, larger machines are normally designed to be repaired.
4) Is the machine economical to repair? Just because a machine can be repaired doesn't always mean that it is also economical to repair. If you have decided your machine is cheaper to repair than replace, then list all subcomponents and consumables that are likely to be required during a normal repair.
5) Next ask: Does the manufacturer have subcomponents and consumables on hand?

Figure 4.2 A small, mag-drive centrifugal pump as shown here may be an example of a machine that is not economical to repair in-house. The user may decide to simply stock a complete unit in stores.

a. If the answer is yes, then you don't need to stock any spare part or consumables.
b. If the answer is no, then you need to stock any spare part or consumables that are likely needed during routine repairs. If you are not sure which to stock, talk to the original equipment manufacturer (OEM).
c. Every time consumables are removed from stores they must be replaced.
d. Every time a spare part is removed from stores, it either must be replaced in-kind part or renewed (see guidelines for *renewable* subcomponents below) and then returned to stores. If the time to refurbish the spare is less than the mean time between repairs (MTBR), then an A-B sparing system, where "A" is the initial part installed in a machine and "B" is a spare in the warehouse, is an acceptable philosophy. There may be legitimate reasons to utilize an A-B-C sparing arrangement, where "A" is the installed part, and "B & C" are spares in stores. The first reason is if the time to refurbish is greater than the MTTR, which means it's possible that a spare part removed during a repair may not be repaired and restocked before the next repair. The other reason to use an A-B-C sparing system is if there is a

real possibility a used part may be found to be unusable upon its inspection during a repair. An A-B-C sparing system provides a time cushion, in the event of unforeseen issues and delays.

6) Are any of the machine subcomponents that are likely to be required during a normal repair renewable (see Figure 4.3)?
    a. If the answer is yes, then list all subcomponents that are economical to renew.
    b. Can any of the renewable subcomponents be renewed during repair outage?
        i. If the answer is yes, then plan to remove these subcomponents and have then renewed during the outage.
        ii. If the answer is no, then you must have a spare set of renewed subcomponents in stores that are ready for use. During every repair cycle, you will remove the used subcomponents from the machine being repaired and replace them with renewed subcomponents in stores and return the used subcomponents to the stores for refurbishment.

Figure 4.3 Some gas turbine subcomponents in the path of the hot gas, such as interstage nozzles and blading, may be renewable. However, they must first be inspected to ensure their integrity before being refurbished.

Some Stocking Examples

Here are a few spare part stocking examples to illustrate the thought process that goes into developing a spare parts program.

Example #1: Small electric motor (non-critical).

Small electric motors are typically considered throwaway items, i.e., they are uneconomical to repair. Here are some options for machines that are not repairable (small pumps and motors)

1. Stock complete spare unit in stores
2. Buy replacement unit from manufacturer if readily available

Example #2: Medium centrifugal pump (non-critical). We assume that the pump is cheaper to repair than replace. Here are some options for machines with internal parts that are replaceable but are not renewable.

- Stock all replaceable parts that are likely to be needed during a normal repair but are not readily available from the supplier.
- Replace all damaged parts from stores (or from manufacturer) and scrap parts removed during repair, assuming that removed spare parts have no salvage value. Then, restock all parts removed from stores.
- Have consumables on hand if they are not readily available from supplier or order from supplier during outage if time permits

Example #3: Spared centrifugal pumps are installed in a high-pressure, high-temperature process application. Due to the severity of the application, there are times when one of the pumps is in service for less than 2 years between repairs, i.e., MTBR <2 years, which is well below a desired 3- to 4-year repair interval. It is common sense to maintain a higher spare parts stocking level for machinery that is less reliable than for machinery that is more reliable. As an example of stocking levels, consider Table 4.1. First, notice that the replacement parts that are likely to be replaced during an overhaul for this hypothetical pump are highlighted. Items like wear rings, shaft sleeves, mechanical seals, and bearings must always be on hand during the repair of either of the main or spare pump. Also, notice the stocking levels for higher usage parts, shown in Table 4.1, are higher for pumps with MTBR less than 3 years than those with MTBR greater than 3

Table 4.1 Example of stocking levels based on equipment reliability criteria.

Part	**MTBR > 3 to 4 years**	**MTBR << 3 years**
Impeller	1	1
Wear rings	**1 set**	**2-3 sets**
Casing	0	0
Shaft sleeves	**1**	**2-3**
Mechanical seal	**1**	**2-3**
Bearings	**1 set**	**2-3 sets**
Bearing covers	0	0
Baseplate	0	0
Coupling	1	1
Motor	1	1

to 4 years. Factors that will ultimately determine actual stocking levels of a given spare part are a) actual MTBR of a set of spared machines, b) time to acquire the spare part, c) mean time to repair (MTTR), and d) likelihood that the part will be required during a standard repair. If several duplicates of any given part are stocked, make sure their usage is tracked along with each parts history, so you can identify its age and vendor in case problems arise when the part is installed.

Example #4: Spared reciprocating process gas compressor (non-critical) The repair time is sufficient to acquire required consumables and renew major subcomponents. Here are some possible approaches to spare parts management:

- Stock all smaller, replaceable parts that are likely to be needed during a normal repair but are not readily available from the supplier.
- Replace all damaged small parts from stores (or from manufacturer) and scrap parts removed during repair. (This assumes removed spare parts have no salvage value.) Then, restock all parts removed from stores.

- Have consumables on hand if they are not readily available from supplier or order from supplier during outage if time permits.
- Renew major subcomponents during the outage. An example of a renewable subcomponent is a piston rod that can be refurbished and then reinstalled. (See more *renewable subcomponent* definition for more examples.) Since the machine is non-critical, the time for renewal is not a concern.

Example #5: Critical gas turbine driver, i.e., unspared. Due to its criticality, the gas turbine repair time needs to be minimized and the total downtime needs to be limited to the allotted maintenance time window. An extended downtime will result in a large economic loss.

Large rotating machinery (larger pumps, compressors, motors, steam turbines, gas turbines) are typically repairable and have internal parts that are renewable. Here are some guidelines for handling spares for this class of machinery:

- Stock all replaceable spare parts, assemblies, and consumables that are likely to be needed during a normal repair but are not readily available from the supplier.
- Replace unrenewable parts removed during repair with new parts from stores (or from manufacturer). Then, restock unrenewable parts and consumables that were removed from stores.
- During the repair, replace all damaged parts from stores (or from manufacturer) and scrap unrenewable parts removed during repair. (This assumes removed spare parts have no salvage value.) Then, restock all parts removed from stores.
- Renewable parts: either a) remove and repair major renewable parts from stores during outage, or b) have renewed spares in stores that are ready for use. If the renewal time for spare parts is longer than the repair time, then it makes sense to keep renewed spares in stock.
- For critical machines with renewable subcomponents, it makes sense to have two sets of renewable spares: one set in the running machine and another set kept in stores. During a repair, remove and replace renewable parts with those previously repaired and held in stores. When parts are no longer renewable buy new replacement parts for the warehouse.

The warehouse is responsible for the following tasks:

1) Assigning unique stocking numbers to all spare parts and consumables.
2) Defining required quantities on hand and trigger points for restocking.
3) Tracking all spare part quantities and statuses.
4) Tracking renewable items that are out for repairs. Critical renewable subcomponents must be returned before the next repair.
5) Determining when renewable subcomponents should be retired. For example, reciprocating compressor valves must be retired after a given number of repairs. When the subcomponent must be retired, then a replacement subcomponent must be reordered, and the retired item must be disposed of.
6) Replacing or eliminating parts that have been made obsolete due to redesigns or retirement of machinery
7) Keeping a list of approved vendors for parts, repairs, and consumables
8) Preserving capital spares by:
    a. Maintaining oil mist systems used for preservation
    b. Inspecting coatings and wraps used for preservation
    c. Rolling rotors on a regular basis
    d. Tracking shelf lives

Capital Spares

Capital spares (defined in *definitions*) are expensive, long-delivery items that may be required for a critical machinery repair that could potentially lead to a prolonged shutdown if unavailable and in the worst cases potentially put you out of business. Normally, few of these items are required during the lifetime of the equipment. Hence, management must decide which parts and how many parts should be held in stores. Money may be saved if capital spares are purchased at the time machines are purchased, since the manufacturer can cut costs by making multiple parts at the same time. (*Sibling Spare* is defined in definition section.) An economic analysis should be made to determine which major spare parts require capital spares.

Since capital spares may be depreciated, they should not be treated as normal inventory. While each site may define the exact criteria for capital spares, it is suggested that spares should be separated using the 80/20 principle for their value classification. Those that have a life span of more than 1 year may be capitalized and depreciated accordingly, while those with one-off uses and little value, and with a life span of less than a year, should be expensed.

Insurance Spares

An insurance spare is any rarely used, long-delivery, spare subcomponent for a machine whose penalty cost for downtime is extremely high and is required to be kept in store by the insurer. Therefore, by definition, they represent insurance against catastrophic failures for which the consequences of an extended down time are very high. These parts do not become obsolete until the parent equipment is retired from service, even if they do not move for many years.

A centrifugal compressor rotating assemble (see Figure 4.4) and a reciprocating compressor crankshaft are examples of insurance spares. An economic analysis is often required to justify insurance spares. Here is a simple example: the cost a spare centrifugal compressor rotating assembly is $1,500,000. Let's assume a) there may be a compressor wreck every 20 years, b) a replacement rotor has a delivery of 3 months, and

Figure 4.4 A complete rotating assembly for a centrifugal compressor, as shown here, is an example of a capital spare. It makes sense in critical services to have a spare rotating assembly in stores in the event of severe rotor damage caused by a catastrophic compressor failure.

c) the economic loss due of downtime is $250,000 per day. Therefore, the annualized value of downtime is equal to $250,000/day*90 days/20 years=$1,125,000/year. (We can also say that the annualized risk related to a catastrophic failure is $1,125,000/yr.) The payout in this simple example would be $1,500,000/$1,125,000 = 1.333 years. In this case, it makes sense to purchase the spare rotor for insurance.

Analyzing Spare Part Inventories Using Monte Carlo Simulations

A Monte Carlo simulation is an advanced analysis tool that can be used to optimize spare part stocking levels for a machine or group of machinery. This tool can be used to determine initial spare part stocking levels to modifications to the existing spare part stocking levels. First, statistical component failure and repair data are entered for each machine in the model. Then, multiple analytical simulations are run by the software to obtain a theoretical distribution of spare parts usage and time between system outages. The power of a Monte Carlo simulation comes from running hundreds or thousands of simulations in order to obtain statistically significant usage and outage distributions that can be analyzed. Various simulation results can then be used to determine the effect of changes on spare parts inventory on the main time between unit outages.

Closing

An ineffective spare parts management program can negatively impact a site's machinery reliability. For this reason, spare parts management should be an integral part of every machinery reliability program. However, often spare parts management tends to fly below the radar and is often given little thought or consideration. To maintain the effectiveness of the program, I recommend that a team consisting of a rotating equipment engineer, shop foreman, warehouse specialist and a management representative be involved in all major spare parts decision and tracking. The team should also meet regularly to discuss: 1) the status of all critical spares, 2) current spare part stocking levels, 3) changes in stocking levels, 4) decisions to dispose of obsolete inventory, and 5) addition or deletions of suppliers, etc. After some trial and error, an optimal mix of machinery spares will be determined for your given machinery population.

Some Definitions Related to Spare Parts

Capital Spare: Any costly, long-delivery item that has the potential to significantly prolong a machinery repair if it is unavailable during the repair sequence.
Component: A complete assembly of subcomponents, i.e., smaller components, that make up a larger system, i.e., pump, compressors, etc.
Consumables: A machinery part that can only be used once then must be discarded, such as gaskets and bolts.
Insurance Spare: Any costly, long-delivery part that an insurance company requires to be purchased and kept on-site. Compare to Capital Spares.
Maximum number of refurbishments: The most times a part can be renewed before it must be scrapped due to loss of critical dimension or mechanical integrity.
MTBF: Mean time between failures.
MTBR: Mean time between repairs.
MTTR: Mean time to repair.
Planned (or scheduled) maintenance: Repairs are scheduled based on running hours or calendar days.
Renewable subcomponent: A machinery part that can be restored economically to a serviceable condition. Some examples of repairable machinery parts are: 1. Mechanical seals, 2. Reciprocating compressor valves, 3. Shafts with wear surfaces, 4. Packing boxes (for steam turbines and reciprocating compressors), 5. Gas turbine nozzles, 6. Gas turbine blades, 7. Control valves, and 8. Relief valves.
Retirement: When a part, assembly, or complete machine has been repaired multiple times and can no longer be returned to manufacturer's specifications it must be scrapped. Parts and assemblies, such as reciprocating valves, transition pieces, and packing boxes, can only be refurbished a limited number of times before they must be discarded.
Routine repair: A normal repair when a machine has reached the end of its useful life.
Routine spares: A group of subcomponents assigned to a given machine that is kept in stores for normal repairs.
Sibling spares: Spare parts purchased when the machine was purchased. These parts are Identical because they were made by the manufacturer at the same time as those installed in the machine were made.
Spare assemblies: A set of smaller parts that are normally assembled and stored as a set, such as rotors, mechanical seals, and reciprocating compressor valves.

Spare Part: Any machinery component or assembly kept in stores until it is needed for a repair.
Subcomponent: A part that makes up a larger more complex component, i.e., impeller, shaft, shaft seal, etc. Hierarchy: Subcomponents→Components→System
System: An arrangement of multiple components designed to work together to perform a useful function. Examples: a processing unit with multiple process machinery, interconnecting pipes, and vessels or a pipeline system with multiple pumps and interconnecting piping.
Unplanned maintenance: An unexpected failure that results in an unexpected repair.
Unrepairable component: Any machinery part that cannot be refurbished or is uneconomical to restore to a serviceable condition.

5

Switch-Over Methodology and Frequency Optimization for Plant Machinery

By Abdulrahman Alkhowaiter

Figure 5.1 Example of a two out of three (2oo3) centrifugal pump installation in the field.

Industrial plants utilize spared machinery to provide higher plant availability as machines can fail and standby machinery is critical to plant reliability and safety (Figure 5.1). This means that plants incorporate a scheduled switchover frequency usually of a one-month period between the operating machine and its standby sister equipment. Some plants use intervals of two weeks, others may use a two-month interval.

Robert X. Perez (ed.) Maintenance, Reliability and Troubleshooting in Rotating Machinery, (75–86) © 2022 Scrivener Publishing LLC

This report attempts to optimize the switchover period to maximize plant machinery safety, reliability and reduce operator workload. The author has worked with over 200 process plants during the past 25 years, and over time recognized that Industry in general is not applying reliability-centered maintenance towards switchover practices of standby equipment.

Machinery Switchover Frequency Optimization Benefits

- Prolong Machinery life and MTBF including the auxiliaries.
- Reduce emergency work activities. Increase machinery and plant safety.
- Minimize operating expenses and total maintenance cost including PM.
- Minimize machinery component corrosion, crystallization of fluids, and adhesive bonding.
- Eliminate rotor sagging and vibration problems.
- Establish true "Hot Standby" capability by actual system testing.

Time-Dependent Issues Involved in Setting Switchover Frequency for Standby Machines

1. **Machinery shafts and rotors** experience some deflection due to gravity weight while the driver and driven machines are stopped. Shaft sag is a time-dependent phenomena impacted by rotor weight and stiffness. It imparts a temporary bend in rotors which causes vibration and possible rubbing problems. This is important particularly for larger machines of 500 KW and above which have greater mass and rotor span between bearings. It is not practical to ask plant personnel to perform hand rotation of all shafts to mitigate shaft bending.
2. **Bearing Housing Component Corrosion:** This occurs to machinery such as pumps and compressors, gearboxes,

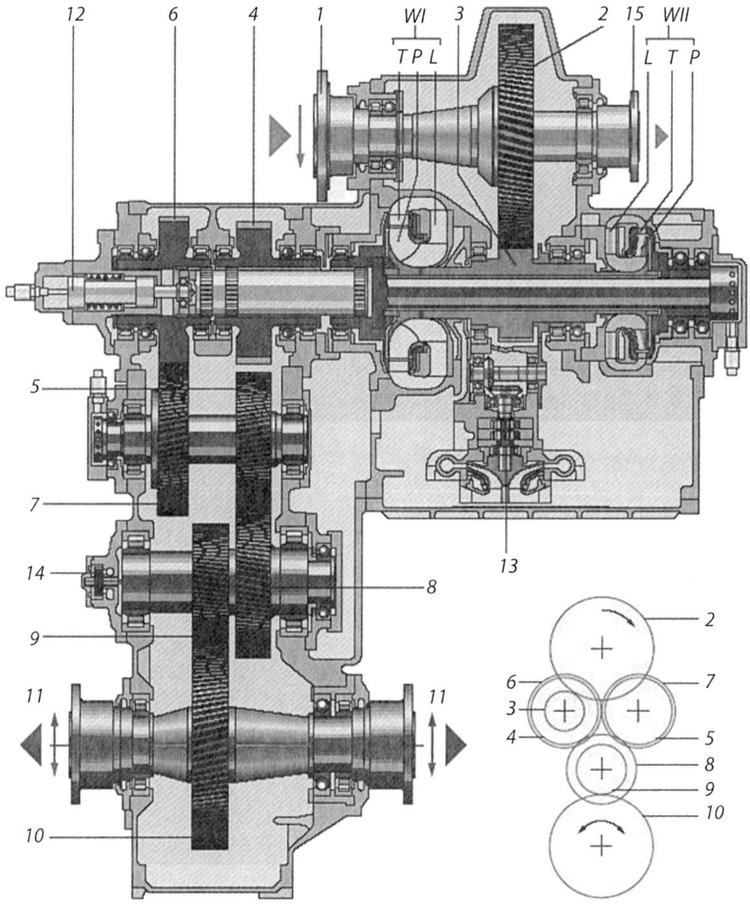

Figure 5.2 Voith germany gailroad gearbox.

motors, and their bearings depending on design, including the shaft sealing areas affected by air humidity while shut down. Both sleeve and rolling element type bearings must be **lubricated once monthly [max interval] or surface corrosion** occurs to the shaft or bearing. In addition, gearbox teeth are made of alloy steel, and must be lubricated with a film of oil at least once monthly, or corrosion attack from the air will damage tooth surfaces (see Figure 5.2).

3. **Centrifugal Compressor:** Dry Gas Seal component materials are chosen from extremely corrosion-resistant materials, so no issues arise with these elements. However, it is

recommended to maintain the nitrogen purge on tertiary DGS seals, functioning all of the time and this will reach the secondary seals at least. The Nitrogen purge pressure can be reduced to a lower pressure such as 2.0 psig during the long shutdown period. This reduces consumption.

4. **Pump Mechanical Seals:** These liquid seals need to have the sleeve and faces rotated with fluid to reduce fouling buildup, adhesion and "lubricate" contacting parts. The adhesion force during startup will cause face and spring failures upon startup after longer time intervals.
5. **Internal Corrosion or Solid Fouling:** In some process machinery applications these may have internal corrosion action due to air or gases trapped in the casing on long shutdowns. This includes reciprocating engines and gearboxes whose gears require lubrication of all surfaces at least once monthly to protect from oxygen and moisture corrosion in air. In some pumping applications, liquids may be crystalizing or fouling, which is another reason to start them at monthly intervals in order to minimize fouling buildup and unbalance.
6. **Rolling Element Bearing Static Load Degradation:** It has been claimed that machinery rolling element bearings undergo microscopic deformations due to the dead weight of the rotor when machine is off. The author has not seen evidence for this in 25 years of plant operations. However, it is possible that this was a phenomenon of the past when rolling element bearings were underrated by design, meaning undersized compared to modern standards, thus resulting in excessive loading under static and dynamic weight. This may have been true for the 1960s-era machinery and before, but not modern machinery. Another reason why this may have occurred long ago is that the precision of manufacturing balls, rollers, and the races they contact was not as geometrically precise as in modern bearings. Modern bearings actual ball or roller contact area to race is higher, leading to reduced loading per unit contact area. The final difference is that modern steel alloy rolling element bearings utilize Rockwell-C-60 surface hardness and above, while 1960s bearings were about RC-55.
7. **Failure to Start on Demand:** The plant operations personnel need to test the standby machinery to determine if

they have the "hot" standby capability of being immediately started in case of failure of the primary operating machine. This is another reason for the typical once monthly switchover frequency and is a valid argument. However, the confirmation of standby machine readiness and fitness status can be achieved through different methods.

8. **Preventive Maintenance:** Another reason for scheduled shutdown of the main operating process machine is to be able to perform detailed preventive maintenance procedures to the driver and driven machines plus associated electrical, lubrication systems, control, and instrumentation. This includes lubrication replacement for machines with self-contained lube oil bearing housings. Maximum recommended oil replacement interval is six months.

Frequent Switchover Introduces the Following Negative Impact to Rotating Equipment

The machinery lifetime operation chart (Figure 5.3) illustrates the various design, installation, maintenance and operational regimes and their impact

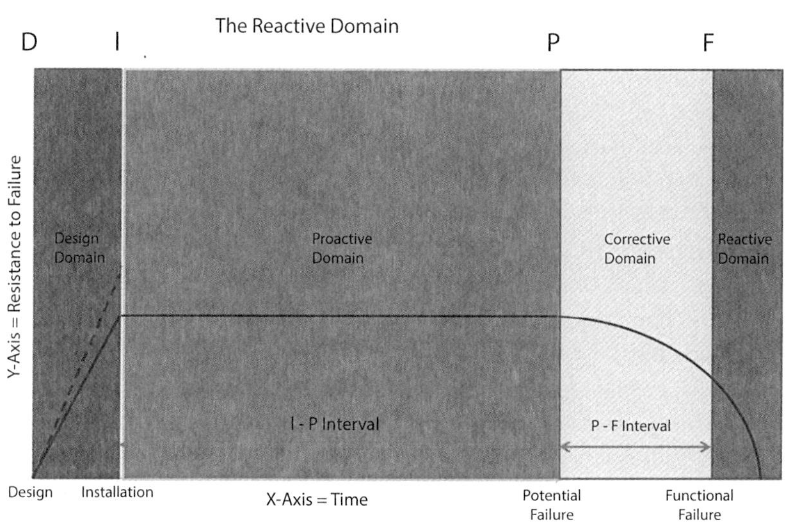

Figure 5.3 Design to installation to potential failure to failure (D-I-P-F) curve. (Source: ReliabilityWeb.com).

on machinery. The failure modes below negatively impact the Proactive and Corrective domains thus reducing total machine life between failures:

- **Pump or Compressor Mechanical Seal Premature Failure:** High startup acceleration torque occurs with motor drive and causes fatigue failures in mechanical seal components on pumps. This is invisible to the user, but it is highly detrimental to seal life, including the thermal cycling failure mode that also occurs at every start-shutdown cycle. Its impact affects liquid seals more than gas seals. For sealless pumps, these are also impacted by thermal cycling mode, and by the possibility of short dry run periods during startups.
- **Internal Rotor Deflection at Startup:** Frequent starting of larger and high-speed machines introduces the hazard of rubbing at close clearance contact zones as the machines pass through critical speeds. For multistage pumps this would be at the interstage bushings and wear rings, while in centrifugal compressors and steam turbines this can occur at packing glands or interstage seals.
- **Process Machinery Infant Mortality:** Introduction of Infant mortality failures due to operator or auxiliary errors at startup affects all machinery. This is a common phenomenon in Fluid handling equipment. For example, it is possible that poor centrifugal pump priming occurs, so each startup introduces the possibility of dry running hazard.
- **Increased safety hazards:** Fires or destructive failure incidents at faulty startup events. It is also possible that fires occur during operation of high-temperature equipment, but what is meant here is that frequent stopping and starting introduces new chances for such failures.
- **Thermal cycling from cold to hot and back to cold:** Acts to fatigue metal, ceramic, and elastomeric parts and loosen bolted components. Thermal cycling affects M-Seals, motors, gas turbines, reciprocating engines, steam turbines, hot service pumps, and gas compressors.
- **Motor Starting Electrical Insulation:** Excess startups reduce motor electrical insulation life particularly on larger motors, with high amperage during startup and sudden high winding temperatures occurring at startup. Impacts both Induction and Synchronous motors.

The main detrimental issues from the above are: sudden acceleration torque at startups, rotor high deflection at startup, thermal stresses at startup and shutdowns, Infant mortality failure at startup, and electrical startup stresses in electrical equipment. The practical conclusion from this is that 75% of damage occurs from starting, with about 25% impact from shutting down cycles. For these reasons, machinery switchovers should be minimized as they are detrimental to reliability. In other words, the standby machine has increased the system availability, but the installation of the second machine has actually reduced the reliability of the first unit. This is non-intuitive, and this interesting conclusion is proven by both theory and actual practice.

Calculation of Start-Stop Damaging Cycles for A, B Configured Equipment: See Definitions Below for More Information

A, B Machines Switching on 1-Month Basis in 12-Month Period
A Machine: 6 starts and 6 shutdowns yearly
B Machine: 6 starts and 6 shutdowns yearly
Total harmful cycles: 24 cycles per year.

A, B Machines Switching on 6-Month Basis in 12-Month Period
A Machine: 1 start and 1 shutdown first six months.
During second half of year 6-month shutdown period: 1 short start-stop per month = 5 cycles.
B Machine: 5 start-stops first half of year, and 1 start, 1 shutdown second half of year.
Total harmful cycles: 7 per machine = 14 cycles per year. Short Stop is not counted as cycle.

A, B Machines Switching on 12-Month Basis in 12-Month Period
A Machine: 1 start and 1 shutdown in 12 months.
B Machine: 11 short start-stops each month during the year. Short Stop is not counted as cycle.
Total harmful cycles: 2 for A-machine, and 11 for B-machine = 13 cycles per year.

Optimum Interval Selection: The author considers that the winner in the above calculations is that which provides the highest balance toward

achieving PM access to machinery, standby active test ability, and greatest reduction of start-stop cycles on main and standby machinery.
This leaves the Six-Month starting-stop model as the ideal.

A, B, C Machines Switching on 6-Month Basis in 12-Month Period
In this case when 3 x 50% machines are installed and 2oo3 are operating:
A Machine: 1 start and 1 shutdown first six months. Then restarted after P.M.
B Machine: 1 start and 1 shutdown first six months. Then restarted after P.M.
C Machine: Is not designated as Standby, it is designated as temporary service machine to be used whenever A or B is shut down during a forced trip or shut down for scheduled maintenance.

Conclusion: For A, B, C service machinery: The third machine will not be utilized as standby switchover machine; it is permanently designated as temporary usage machine and will accumulate less hours compared to sister machines. This machine is started once monthly for one hour. This total procedure is not harmful to machine or plant.

A, B, C, D Machines Switching on 6-Month Basis in 12-Month Period
In this case when 4 x 33% machines are installed and 3oo4 are operating
A Machine: 1 start and 1 shutdown first six months. Then restarted after P.M.
B Machine: 1 start and 1 shutdown first six months. Then restarted after P.M.
C Machine: 1 start and 1 shutdown first six months. Then restarted after P.M.
D machine: Is not designated as Standby, it is designated as temporary service machine to be used whenever A, B, C machine is shut down during a forced trip or for scheduled maintenance.

Conclusion: For A, B, C, D service machinery: The fourth machine will not be utilized as standby switchover machine; it is designated as temporary usage machine and will accumulate less hours compared to sister machines. This machine is started once monthly for one hour.

Definitions

Short Start-Stop: A predetermined scheduled frequency of starting a machine; both driver and driven solely for the purpose of establishing ability to start and function properly with freedom from mechanical, electrical, and system process problems. It has the added function of lubricating parts in the available bearing housing grease or oil lubrication and reduces shaft deflection that occurs from long periods of standstill mode. For example, the machine is started once monthly for 30 minutes, in recycle mode where it is not brought into actual plant operation or production as its sister machines are functioning at that time. One-month interval is optimal.

Definition of Short Stop: This defines an activity of a machine which has undergone a start cycling as explained above and is then shut down after a short period, for example, after 30 minutes of testing its function and establishing reliability and availability. This is not counted as a cycle by itself, as the Short start-stop is the complete cycle. In normal operated equipment, for example operating for one month before switchover, they undergo two cycles, a start cycle and a shutdown or stop cycle. However, for those machines that are only in a short stop-start, such as 30 minutes total, the start and stop are modeled as a single cycle. This is because there was not enough time to cause significant thermal cycling.

Examples of Short Start-Stop Intervals in Process Machinery

1. For reciprocating gas compressors: these can be short stop-started in unloaded condition and then shut down or operated in recycle for a short period.
2. For centrifugal pumps: the standby unit can be operated alone without bringing into service, by short start-stop with 50% capacity bypass or recycle line. 100% bypass sizing not required.
3. For centrifugal or Screw air compressors: these can be discharged to atmosphere, offline.
4. For centrifugal process gas compressors: these are rarely spared, but if in A,B configuration, they are always built with 100% capacity full recycle lines.
5. No harmful effects arise from short start-stop operation mode. All the above machines are not brought on-stream into the process during short start-stop cycling.

Philosophy of Reliability-Centered Switchover Strategy

- ❖ The above calculation attempts to capture the relationship between time interval in operating and shutdown modes, as compared to frequency of damaging cycles.
- ❖ The author considers that the winner in these calculations is that which provides the highest balance toward achieving Preventive Maintenance PM Access to machinery standby active ability and reducing start-stop cycles on both main and standby machinery. **This leaves the Six-Month starting model as the ideal for all types of plants machinery.**
- ❖ This model also incorporates once monthly short start-stopping cycles on standby machines to maintain lubrication of parts, to test total mechanical-electrical-process system readiness and reduction of fouling-corrosion in process machines.
- ❖ Reduction of Machine Stress Monthly Test Starting: In the 6-Month model, the yearly number of starts is the same as for 1-Monthly switchover interval; however, the operating temperatures of critical parts do not reach the higher temperatures achieved under long-term operation. This is why stops are ignored. In addition, during the monthly test startup of reciprocating engine machinery, steam, or gas turbine driven machines, the shaft speed does not have to reach 100%; even 50% is suitable as long as critical speeds are avoided.
- ❖ Improved Operational Philosophy: Maintain the operating blowers, compressors or pumps designated as main machine. The standby compressors or pumps are only started once monthly for one hour on recycle mode then shut down without bringing them online. The standby machine is later placed into full operation at six-month intervals. For A,B,C or A,B, C,D configurations, where the non-utilized machine is not a Standby, it is considered to be a temporary usage machine.
- ❖ There is no benefit to operating centrifugal machines on short interval shared time basis as only positive displacement machines undergo significant wear during operation. Even reciprocating machines can have sharing of wear life in extended intervals of six months. Well-designed and maintained modern centrifugal machines do not need to

be operated on a frequent switching on-off cycle mode of operation.

- Large rotating equipment are equipped with pressure fed circulating lube oil systems, and therefore it is recommended to start these auxiliary systems every week for a 30-minute time period as a system test. This will also eliminate the possibility of corrosion in lube and seal oil systems, including bearing and seal journal areas on rotor. All non-operating pumps or compressors with lube oil systems should have the lube oil system started once weekly for 30 minutes. This weekly action can be safely automated in the plant central control computer as no startup hazards occur from remote starting of the lubrication pumps.
- Preventive Maintenance intervals of self-contained bearing housings can be extended by using high-quality synthetic oils which allow longer operation without oil replacement.
- This method reduces startup-shutdown cycles per year from 24 to 14 for the machinery train assuming A, B, or more configuration machines that were normally switched once monthly. At the same time, it confirms that the standby machines are available for action.

Part II

DESIGN AUDITS AND IMPROVEMENT IDEAS

6

Evaluating Centrifugal Pumps in Petrochemical Applications

By Robert X. Perez

Figure 6.1 Cutaway of an axially split, multistage centrifugal pump. These types of pumps are used whenever a liquid process stream needs to be transferred at a high pressure.

Overview: *The author explains how the processes found in petrochemical facilities result in a wide variety of pumping applications. Each of these applications must be carefully evaluated to ensure that the best pump is chosen for the service. In cases of unreliable pumps, the author covers how to review problematic pump installations to determine if the pump or the pumping system is to blame. Also covered are some of the most challenging pumping applications the author has encountered in the field along with ways to avoid their pitfalls. This brief survey of pump applications will help pump professionals better assess, select, and troubleshoot pumps in their facilities.*

Robert X. Perez (ed.) Maintenance, Reliability and Troubleshooting in Rotating Machinery, (89–140)

There may be several reasons for evaluating a pump (see Figure 6.1) in your petrochemical facility. The pump or pumps may be failing frequently, not performing as expected, or vibrating or pulsating excessively. The evaluator wants to know: Is there a pump design issue, operating issue, or pump condition issue? Before beginning, the auditor must realize that there is no way to divorce the pump from the pumping system. The auditor should always think about the pumping system as an interconnected assembly made up of the suction piping systems, discharge piping system, pump, shaft sealing system, driver, control system(s), and lubrication systems. All these subsystems must be matched properly with the pump to ensure reliable performance. If each subsystem is carefully designed and matched to the pump, then there is a high probability that the pump will work reliably. Furthermore, the process must provide a steady source of liquid to the pump and the composition of supply liquid must be consistent and fall within the design specifications of the pump.

Before beginning a field pump review, here are a few issues to consider:

- Is the pump spared or unspared? Criticality is an important consideration when determining the reliability requirements for the application.

    - When dealing with spared pumps, the mean time between failure (MTBF) metric is an accepted measure of reliability; however, with unspared pumps, users are more concerned about the probability of operating continuously from one process outage to the next.
    - If the pump is fully spared, compare the mean time between failures to similar pumps in the facilities and in other petrochemical facilities. A thorough review of the pump and installation is needed if the failure rate is found to be excessive.
    - If the pump is unspared, determine if there are any instances where it failed prematurely and resulted in a unit or plant shutdown.

- What is the operating temperature of the product being pumped?

    - Pumps operating at temperatures >350 °F should meet API design requirements,

- Higher temperatures lead to piping expansion, alignment issues, and pump growth issues; so, make sure piping around hot pumps has been designed to allow for expansion to limit nozzle loads.
- Higher temperatures require special seals and flush plans. Following the API 682 guide is highly recommended.

- What is the vapor pressure of the product being pumped at the operating temperature? If the pressure drops below the liquid's vapor pressure inside the pump, cavitation will occur. Remember that pure products, such as ethane, propane, butane, and water are more volatile than hydrocarbon mixtures, so they must be handled with more care.
- What is the net positive suction head (NPSH) margin? If the NPSH margin is marginal, then small changes in the operating conditions could lead to sufficient pump cavitation to affect performance or cause impeller damage.
- What is the flowrate as a percentage of the BEP (best efficiency point)? Operating pumps away from their design point can lead to internal recirculation, low efficiencies, high radial loads, and vibration, which could lead to early failures.
- Are you pumping a high viscosity product? A high viscosity product could 1) affect pump performance, 2) require special sealing configurations, and 3) require special operating procedures.
- Are there solids in the product? Slurry pumps, Coker pumps, Bottoms pumps, and trash pumps require special designs to ensure reliable performance.
- Is there enough settling time in the suction vessel to allow vapors to disengage? Vapor entrainment can appear to be cavitation. Adding more surge volume or settling weirs can help promote vapor disengagement and prevent two-phase flow conditions in the pump suction line.
- Are there any piping or valve issues?

  - Look for suction piping issues, such as high points, or undersized piping. Field experience has shown that the ideal minimum length of full-size, straight pipe immediately upstream of the suction nozzle is 10D (10 pipe diameters) and the absolute minimum length of full-size, straight should be 5D.

- Look for control valves that are operating out of range, i.e., almost fully closed. Flow control valves that operate almost fully closed may indicate that a pump rerate opportunity exists. A control valve that operates almost fully closed is an indication that there is too much head being generated by the pump, which means you may be able to trim its impeller or de-stage it.

- What is the pump's control scheme? Is it controlled based on a tower level, the net flow, or on a minimum flow spillback? A pump problem could actually turn out to be a control problem.
- Has the pump experienced process upsets in the recent past? A low level can result in pump cavitation and temperature excursions will lead to seal failures.
- What is the seal flush plan? Does the existing seal fluid plan provide the required level of reliability for the application? Compare the seal rate with the plant average and with other seals in similar applications.
- What is the monitoring strategy for the pump? Is the pump remotely monitored? Is the pump checked on operator rounds? Are there local vibration monitors? Are there seal alarms? Are there flow alarms that can be seen in the control room?
- Are there operating procedures for startup, upsets, shutdowns? Improper operating procedures can significantly affect pump reliability.

We know that pump applications vary greatly across process facilities. There is no such thing as a typical pump in the petrochemical industry. Pumps come in a wide range of sizes, metallurgy, and configurations, and can have various auxiliary systems, such as lubrication and seal flush support systems. To better understand the various pump applications found in petrochemical facilities, we first need to understand a little bit about oil refining and the various processes associated with hydrocarbon processing.

Crude Oil Processing

Petroleum refineries are large-scale industrial complexes that produce marketable petroleum products from crude oil. U.S. refineries primarily focus on the production of gasoline due to its high demand, while diesel fuel is the principal refinery product in most other parts of the world [1]. Table 6.1 shows a typical distribution of products obtained from a barrel of crude oil in a U.S. refinery.

Complex hydrocarbon mixtures, such as crude oil, or other petroleum products, are composed of hundreds or even thousands of different compounds with varying molecular weights. For this reason, hydrocarbon mixtures tend to boil over a temperature range, as opposed to having a single point as in pure compounds. The boiling range covers a temperature interval from the initial boiling point (IBP), defined as the temperature at which the first drop of distillation product is obtained, to a final boiling point, or endpoint (EP) after the highest-boiling compounds have evaporated. For example, the boiling range for crude oil can range from 300 °F to 1000 °F. To understand distillation processes, it is important to understand that pure liquids, such as butane and propane, have a specific boiling point, while hydrocarbon mixtures boil over a range of temperatures.

The boiling point range nature of hydrocarbon mixtures enables them to be refined through processes that progressively boil off lighter and lighter fractions. These distillation and fractionation processes can be found in all modern petrochemical processing facilities. Therefore, understanding distillation and fractionation fundamentals is a good way to better understand petroleum processing and why they produce finished and intermediate liquids with a wide range of physical properties.

Table 6.1 Percentage of finished products derived from a typical barrel of crude oil.

Product	Percentage of saleable products recovered from a typical barrel of crude
Gasoline	47%
Diesel Fuel & Hearing Oil	23%
Other Products	18%
Jet Fuel	10%
Liquefied Petroleum (LNG)	4%
Asphalt	3%

It should be noted that not all the gasoline produced in a refinery is obtained from distillation processes. A significant fraction of the gasoline yielded from a barrel of oil comes from catalytic cracking and reformation processes: catalytic cracking is a process where a catalyst is used to convert a substantial part (>50%) of a high-molecular weight hydrocarbon feedstock into gasoline and lower boiling products, usually in a single pass; catalytic reforming is a chemical process used to convert petroleum refinery naphthas distilled from crude oil into high-octane liquid products called reformates. While these processes are vital to refinery performance, review of these catalyst-based processes are beyond the scope of this brief overview of crude processing.

Desalting

Desalting is the first separation process that takes place at the front end of a petroleum refinery (i.e., prior to atmospheric distillation); the purpose of desalting is to prevent corrosion and fouling of downstream lines and equipment by reducing the oil's salt content significantly. Desalting is normally considered a part of the crude distillation unit since heat from some of the streams in the atmospheric distillation is used to heat the crude in the desalting process.

After desalting, the crude is ready for distillation. The distillation process separates the crude oil into various boiling point fractions. (A fraction is any component of liquid mixture resulting from distillation.) The higher boiling fractions lead to most desirable distillate liquids, such as gasoline, jet fuel, diesel fuel, and fuel oil in the increasing order of boiling points. The residual distillation fraction is composed of asphaltines, which can have specific gravities typically greater than 1. The liquefied petroleum gas (LPG) constitutes the lowest boiling point (most volatile) product found in a refinery.

Crude Oil Distillation

Most crude oil refineries employ a separation process called *distillation* to "refine" crude oil. Distillation separates crude oil, which is a physical mixture of various hydrocarbon components, into liquid fractions, based on the differences in their physical properties, such as their densities and boiling points. Distillation makes it possible to isolate more than two components of a mixture in a single processing tower. In a distillation column, the "lightest" products, with the lowest boiling points, exit from the top of the column and the "heaviest" products, which are composed of residue composed of lube oil, paraphing wax and asphalt, with the highest boiling points, exit from the bottom of the tower (see Figure 6.2).

After desalting, the crude oil feed is heated and then fed into the atmospheric distillation column (Figure 6.2). After entering the atmospheric distillation tower, the liquid portion of the feed falls to the bottom of the column, while the vaporized portion of the feed rises to a higher level (or tray) of the distillation tower. On the first tray there is, 1) condensed liquid on the tray which can be drawn off, 2) heavier liquid that falls below the tray, and 3) vapor from below that bubbles through the liquid on the tray. The vapors traveling upwards through the trays' strip lighter ends and carry them upwards. Most trays have many holes or bubble caps (Figure 6.3) attached to them that allow the vapor to pass through, which increases the contact time between the vapor and the liquids in the column and help to collect liquids that condense on the tray.

This first split results in a difference in density between the "bottom" fraction and the remaining liquid in the next higher tray. The fraction at the next higher tray is slightly lighter and cooler than the liquid at the

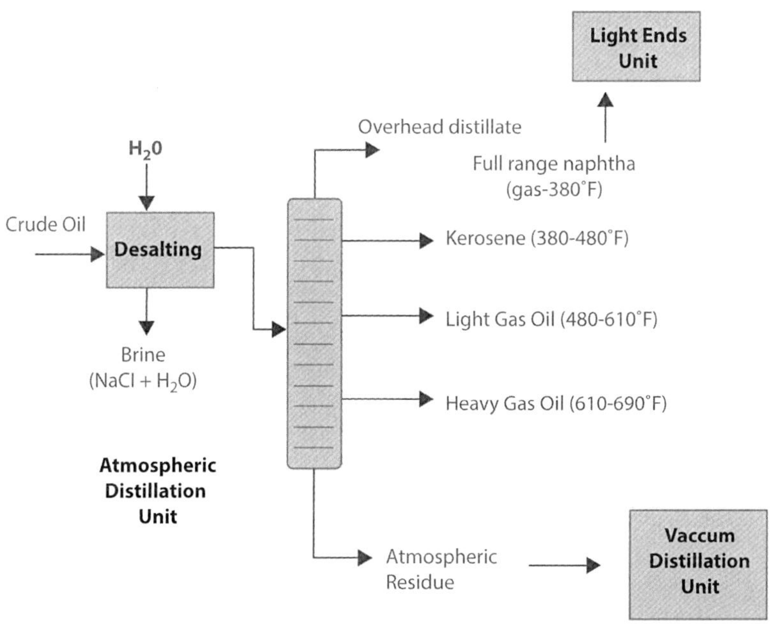

Figure 6.2 A typical atmospheric distillation column. Notice that the heaviest faction (atmospheric residue) exists in the bottom of the column and the lightest fraction exits the top of the column (overhead distillate).

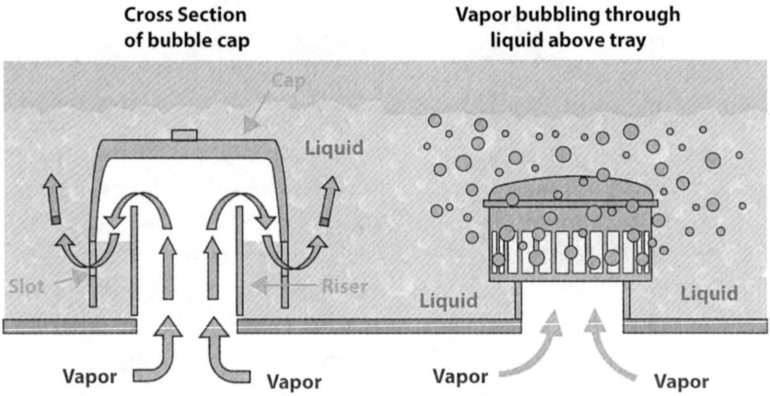

Figure 6.3 Typical bubble caps used on distillation trays.

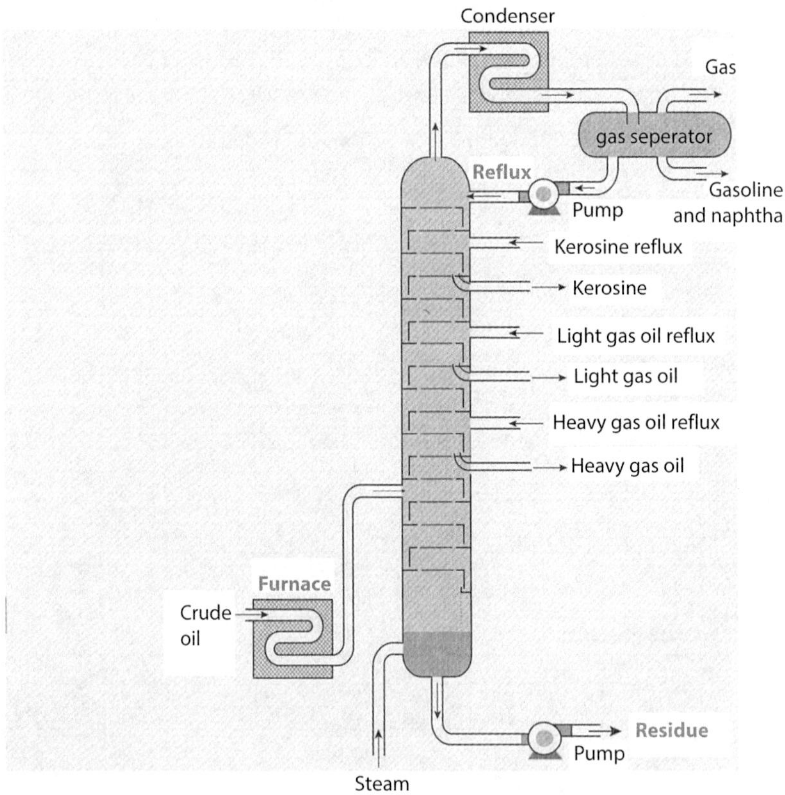

Figure 6.4 Hydrocarbon vapors that exit the top of the distillation tower are cooled with a condenser before entering a gas separator. Some of the condensed liquids collected in the gas separator are returned to the tower as reflux.

bottom of the column. Some of the liquid on each tray can be drawn off and pumped away from the tower or boiled off and travel to the next higher tray. Similarly, the fraction at the second tray is slightly lighter and cooler than the liquid on the first tray.

On each tray, some liquid is drawn off, some liquid boils off, and some of the vapor rises to the next tray. Therefore, the liquid at each higher level gets lighter, and lighter has a lower average molecular weight and density. When vapors reach the top of the tower they are cooled off with a heat exchanger, called a condenser, and partially condensed (Figure 6.4). The condensed liquid, which is pumped back into the top of the tower, is called reflux. (A reflux stream provides a means of establishing a liquid-vapor equilibrium at the top of the column so that the purity of the overhead vapor can be set and maintained.) Any uncondensed vapor is piped away to a vapor processing and collection system.

The temperature distribution seen traveling up the distillation column is controlled by the temperature of the feed into the column and the amount of cooling provided by the reflux flow at the top of the column. Processing is optimized by controlling the feed inlet flow and temperature and reflux flow and temperature. The goal of process controls is to develop a specific

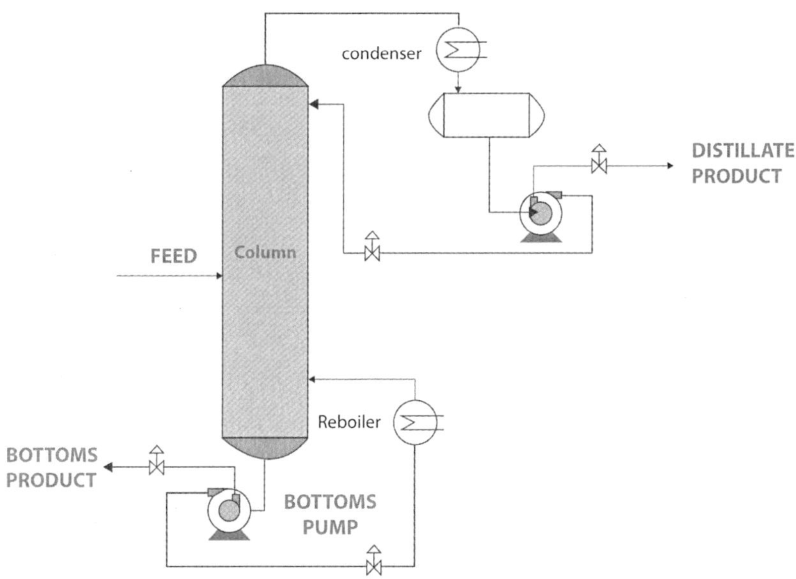

Figure 6.5 Tower with reboiler and reflux circuits.

temperature and density gradient from the bottom to the top of the distillation tower so that desired liquid fractions are separated and extracted for further processing, transfer, or storage. Variations in the composition of the feedstock may require changes in the tower's operating control points.

In distillation processes, heat can be applied to the bottom of the column either by using a feed heater as shown in Figure 6.4 and/or using a reboiler as shown in Figure 6.5. A reboiler system splits the bottoms pump stream into two flows: a) Part of the bottoms flow is heated with a heat exchanger and then reinjected back into the tower, b) the remaining bottoms flow is sent off for further processing or storage.

The basic arrangement of a distillation tower has been revealed: heat is added to the bottom of the column to boil the feed and then cooling is applied to vapors at the top of the column to promote condensation of the lighter fractions. A portion of the condensed light ends are then returned to the top of the column as reflux to provide cooling. Most distillation columns and fractionators use heat and cooling in this way to allow the physical separation of the various liquid fractions to occur in a predictable way within the column.

Properties of Distillation and Fractionator Fractions

Why is it important to understand the properties of distillation fractions? Because each of the various processing fractions will have different physical properties. Physical characteristics, such as density, vapor pressures, viscosity, etc., provide insight that can determine how they can be reliably pumped and transported (Table 6.2). For example, crude bottoms tend to be heavy and have a low vapor pressure. If cooled to room temperature, crude bottoms will become very viscous and could even solidify. In contrast, the liquids condensed from the overhead gases tend to be very light and may be required to be kept under pressure to prevent it from flashing off.

To understand how a liquid will act inside a pump, we need to know the vapor pressure, viscosity, and density at its pumping temperature. As we have noted earlier, we know that liquids in distillation towers and fractionators get lighter as we move up the tower. Lighter means it has a lower density and a smaller molecular weight. Lighter liquids have high vapor pressures. (Vapor pressure is a measure of the tendency of a liquid to

Table 6.2 Typical physical properties of crude oil fractions. Notice that as the boil point temperature of a fraction increases its specific gravity increases, i.e., the fraction is heavier, and its viscosity increases. The specific gravities of crude fractions range from 1.02 for vacuum resid to 0.67 for light naptha.

Crude fraction	Boiling point range (F)	Specific gravity	Viscosity at 104 F (Cst)	Viscosity at 212 F (Cst)
Vacuum Resid	1050	1.02	200000000	20000
Heavy vacuum gas oil	850-1050	0.96	3510	28.1
Light vacuum gas oil	650-850	0.94	39.95	6.41
Atmospheric gas oil	500-650	0.89	4.96	1.66
Kerosene	400-500	0.85	2.21	0.95
Heavy naptha	300-400	0.82	1.35	0.67
Medium Naptha	175-300	0.77	0.87	0.49
Light naptha	50-175	0.67	0.6	0.38

change into the vapor state. The higher the vapor pressure the more volatile the liquid is and the more readily it will evaporate or boil.)

Let's look at a hypothetical vapor pressure curve in Figure 6.6. The red curve is the boundary of liquid and vapor over a given temperature range. Under equilibrium conditions, if a fluid is on the left side of the red line, it will exist as a liquid. Conversely, if the fluid in on the right side of the red line, it will exist as a vapor. If the pressure drops below the vapor pressure curve at a given temperature, flashing with occur. Pumping and sealing problems occur when a process is operating close to the vapor pressure curve. Consider the particle in the figure. If the process fluid is heated to the point it moves to the right and reaches the vapor pressure curve,

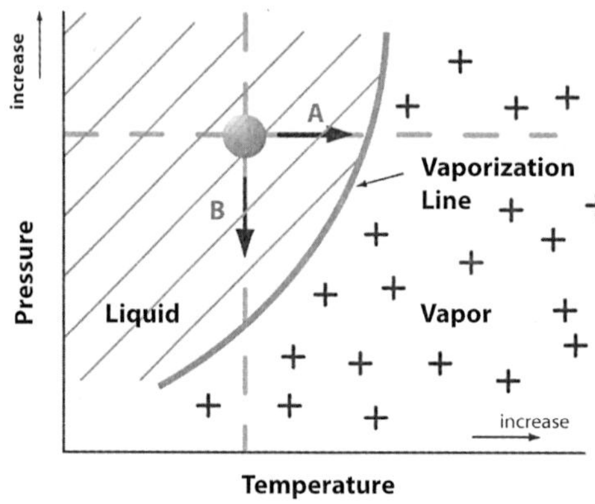

Figure 6.6 The red curve is a vapor pressure curve for a hypothetical liquid. To the left of the red curve, the fluid is in the liquid state and to the right of the red curve the fluid is in the vapor state.

flashing will occur. Another possibility is that the temperature remains the same, but the pressure drops. In this case, the fluid will drop down to the vapor pressure line and flash. Therefore, there are two ways to cause a liquid to flash: increasing the temperature until it reaches the vapor pressure curve (A), or 2) dropping the pressure until it reaches the vapor pressure curve (B).

Pump cavitation is the result of liquid flashing as it enters the pump's suction. Like path "B" seen in Figure 6.6, as liquid enters the pump's suction the pressure will drop depending on the pump's internal design and operating speed. If the liquid's pressure drops below the liquid's vapor pressure, flashing will occur. The difference between the inlet pressure and the lowest pressure inside the impeller is called the net positive suction pressure required ($NPSH_r$). When selecting a pump, always ensure there is sufficient suction head to avoid excessive flashing inside the impeller. In other words, the net positive suction head available ($NPSH_a$) should always be greater that the $NPSH_r$ to prevent a loss of performance or impeller damage. Always try to include a safe margin between the NPSHa and the NPSHr. (See NPSH Margin insert below.)

Defining NPSHr, NPSH3, and NPSH Margin

The net positive suction head required (NPSHr) is the minimum net positive suction head recommended by the pump manufacturer to ensure the pump achieves its specified performance. Most pump manufacturers use the term NPSH3, which is determined during testing, to define NPSHr. NPSH3 is the net positive suction head required to limit the loss of total pump head to 3% loss, due to cavitation. Keep in mind that even before the NPSH3 point is reached, cavitation is already occurring in the impeller; but even though some cavitation is occurring, in most non-critical pump application, using NPSH3 values as a design constraint has been proven to be a good trade-off between pump reliability and life cycle costs.

Remember, all pumps operate with some suction cavitation. The only question is whether the cavitation is sufficient to damage the impeller or noticeably degrade performance. Experienced designers incorporate a safety margin between the NPSH3 of the pump and the NPSHa to avoid levels of cavitation that could affect pump reliability. The Hydraulic Institute offers guidelines for NPSH margins (See "The NPSH Margin" section below). The recommended suction head margins can be based on either the:

1. NPSH Margin Ratio, which is the NPSHa divided by the $NSPH_3$ or
2. Margin which is the NPSHa minus the NPSH3

In this chapter, the term NPSHr is equivalent to the $NPSH_3$ term defined above and vice versa.

Remember that pure liquids, such as butane and propane, have a specific boiling point, while hydrocarbon mixtures boil over a range of temperatures. Why is this important? Liquid purity effects the relative volatility of the liquid being pumped. A pure liquid will boil or flash more violently than a liquid hydrocarbon mixture, since initially only a portion of the liquid making up the fraction flashes. This difference explains why you need to be more careful when pumping pure liquids than liquids mixtures.

Natural Gas Processing: NGL Processing

Other processing facilities commonly found in the petrochemical industry are those designed to purify raw natural gas by removing impurities, contaminants, and higher molecular mass hydrocarbons to produce pipeline-quality, dry natural gas. Once the gas stream has had the acid gases and water vapor removed, it is classified as dry, sweet gas, and is suitable for even further processing.

Today, cryogenic processes, where the temperature of the gas stream is dropped to around -120 °F, are commonly used to extract NGLs from the rich natural gas. In this process, the natural gas stream is first cooled, by sending it through an expansion turbine, also called a turboexpander. The turboexpander rapidly expands the gas stream, which causes the temperature to drop significantly. This rapid temperature drop condenses ethane and other hydrocarbons in the gas stream, while maintaining methane in gaseous form. In addition to the cooling benefits of the turboexpander, its "expander wheel" recovers sufficient shaft horsepower to drive the attached "compressor wheel," which in turn is used to recompress the gaseous methane effluent, thus saving energy costs associated with extracting ethane. This process allows for the recovery of about 90 to 95% of the ethane originally in the gas stream.

Once NGLs have been removed from the natural gas stream, they must be broken down into their base components to be useful. That is, the mixed stream of different NGLs must be separated out. The process used to accomplish this task is called fractionation. (Note: The fractionation is a process used to separate liquids with closer boiling points.) This is done by passing them through a series of fractionation towers to boil off hydrocarbons starting with the lightest to heaviest fractions (see Figure 6.7). Essentially, fractionation occurs in steps consisting of the stages of boiling off heavier and heavier hydrocarbon factions one by one. The name of each fractionator gives an idea as to its purpose, as it is conventionally named for the hydrocarbon that is boiled off. The entire fractionation process is broken down into steps, starting with the removal of the lighter NGLs from the stream. Unit fractionators are used in the following order:

- Deethanizer – A step that separates the ethane from the NGL stream.
- Depropanizer – A subsequent step that separates the propane.
- Debutanizer – Another step that boils off the butanes, leaving the pentanes and heavier hydrocarbons in the NGL stream.
- Butane Splitter or Deisobutanizer – A step that separates the iso and normal butanes.

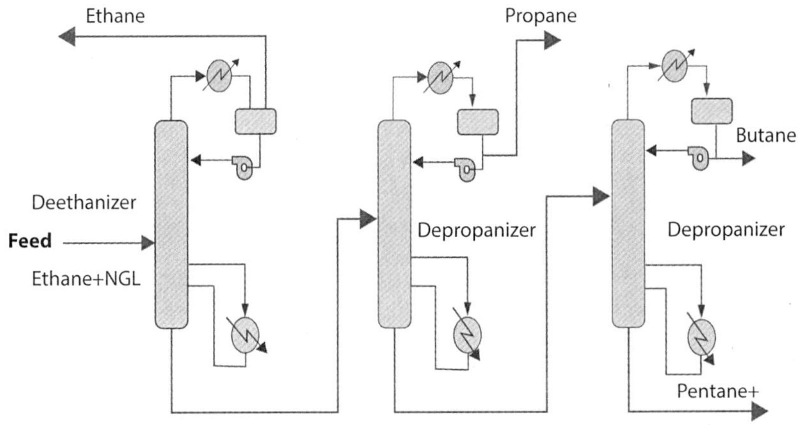

Figure 6.7 Typical fractionation train used to produce three liquid common NGL products.

By proceeding from the lightest hydrocarbons to the heaviest, it is possible to separate the different NGLs efficiently. To maximize the yields of the most valuable products, fractionation processers need the flexibility to not only handle varying compositions of NGLs but also be able to change the mix of hydrocarbon components in the output for sale.

An example fractionation train used to produce three liquid common NGL products is illustrated below (Figure 6.7). The LNG feed stream from the demethanizer (not shown) contains too much ethane to be included in the liquid products; thus, it must be processed in the deethanizer. A portion of the deethanizer's overhead stream is used as reflux and the remaining portion is routed to product storage. The bottom product from this deethanizer column can be marketed as a deethanized liquid NGL product and pipelined to another larger NGL recovery plant. The second column in our arrangement is the depropanizer which produces a specification propane product overhead. The bottom product, a butane-gasoline mixture, is often sold to a pipeline without further processing. However, to produce butane and gasoline products, a third column for the debutanizer is required. A debutanizer separates the iso-butane and normal butane from the natural gasoline or pentane plus product. This separation is controlled to limit the vapor pressure of the natural gasoline. The overhead butane product can be sold as a mixture or an additional column can be used to separate the iso-butane and normal-butane.

As LNG is processed the liquid products extracted become heavier and heavier and their vapor pressures get lower and lower (see Table 6.3).

Table 6.3 Boil points and densities of natural gas fractions.

Name of fraction	Formula	Boiling point (°C)	Density	Physical state (at 20 °C)
Methane	CH_4	-164	0.668 (g/L)	Gas
Ethane	C_2H_6	-89	1.265 (g/L)	Gas
Propane	C_3H_8	-42	1.867 (g/L)	Gas
Butane	C_4H_{10}	-1	2.493 (g/L)	Gas
Pentane	C_5H_{12}	36	0.626 (g/mL)	Liquid
Hexane	C_6H_{14}	69	0.659 (mg/L)	Liquid
Octane	C_8H_{18}	125	0.703 (mg/L)	Liquid
Decane	$C_{10}H_{22}$	174	0.73 (mg/L)	Liquid

This fact is important when selecting pumps to handle these fractions. We will discuss later how a liquid's density has a significant effect on a pump's design, i.e., specific speed, and number of stages, required to generate the required head for a given application.

Centrifugal Pump Design Audits

We have seen how petrochemical processes by their nature require a wide variety of pumps that can operate across a vast range of flows, pressures, temperatures, and viscosities. This diversity of pump designs requires rotating equipment professionals to be well versed in the art of pump design and the common problems encountered in the field. Regardless of the type of the process pump being evaluated, it is good practice to begin the evaluation by reviewing the following aspects of the problematic pump under audit:

- The design standard selected
- The materials of construction
- The hydraulic fit
- The NPSH margin
- Seal and seal flush design

We'll now review these design aspects in some detail to understand why they are important to a pump's reliability.

Design Standards

Before selecting or evaluating centrifugal pumps, users should become familiar with the various quality standards used by pump professionals. In higher pump criticality or challenging applications, higher design and construction standards should be adopted. Here are the most common pump standards used in petrochemical applications:

ANSI (American National Standards Institute) pumps: ANSI standards refer to dimensional standards. ANSI pumps are also required to meet ANSI B73.1 standard, also known as ASME B73.1 standard (American Society of Mechanical Engineers). The objective of this standard is to ensure interchangeability of ANSI process pumps of similar sizes. These centrifugal pumps are available in horizontal, end suction, and single-stage pump configurations, regardless of the manufacturer.

API (American Petroleum Institute) pumps: API standards refer to the parameters of a pump's construction, design, and ability to handle high temperatures and pressures. API 610 specifications cover a wide variety of API types, including API VS4, API VS7, API OH3, API OH2, API OH1, API BB1, API BB2, API BB3, etc., designed. These centrifugal pumps must meet the requirements of the American Petroleum Institute Standard 610 for General Refinery Service.

It is recommended [2] that API 610-compliant pumps be given strong consideration when one or more of these six conditions are reached or exceeded:

- head exceeds 350 feet
- temperature exceeds 300 °F on pipe up to and including 6-inch nominal diameter; alternatively, if the temperature exceeds 350 °F on pipe starting with 8-inch nominal diameter
- pumps with drivers rated more than 100 horsepower (hp), starting at 75 kW and higher
- suction pressures over 75 pounds per square inch gauge (psig)
- flow more than the best efficiency point (BEP) for the pump at issue
- speeds more than 3,600 rpm

DIN (German Institute for Standardization) pumps: DIN 24256 specifications. Centrifugal pumps satisfying these standards are used in installations requiring large flow rates, abnormally high working pressures or very high temperatures.

Table 6.4 Comparison of API and ANSI pump standards.

Feature	API 610	ANSI B73.1
Pressure Ratings	All pressures and temperatures normally encountered in processing facilities	ANSI Class 150 and temperatures up to 500 °F
Pump Casings	ASME Section VIII, Div. 1; Carbon steel or alloy steel required for flammable or toxic services	Ductile iron, carbon steel or allow steel required for flammable or toxic services
Impellers	Single piece castings; must be secured to shaft with a key	May be keyed or threaded to shaft
Wear Rings	Minimum hardness of 400 BHN or a hardness difference of 50	No requirements
Mechanical Seals	Required	No requirements
Shaft Critical Speed	Lateral critical speed greater than 120% of the maximum pump speed	No requirements
Bearings	3-year life required for rolling element bearings; Hydrodynamic and thrust bearings when bearing diameter (mm) x pump rpm exceeds 300,000 or, the pump rated horsepower x pump rpm exceed 2.7×10^6.	2-year life required for rolling element bearings
Casing Support	Centerline support of casing	Foot mounting under casing
Testing	Hydrostatic and performance testing required.	No performance testing required

ISO (International Organization for Standardization) pumps: ISO 2858, 5199 specifications, the international standard ISO 5199 specifies the requirements for class II end suction centrifugal pumps of single-stage, multistage, horizontal, or vertical construction, with any drive and any installation for general application.

Table 6.4 lists some of the main differences between the API and ANSI pump standards. Notice that the API requirements for design and testing exceed those in the ANSI standard in every way. API pumps have been proven to provide safe and reliable service in the most difficult process applications, while ANSI pumps are considered to be an economical pump alternative for less severe chemical processing applications.

The Materials of Construction

Selecting construction materials is key to the success of pumps operating in severe applications. Although the initial cost is normally of paramount importance, the operational costs, replacement costs, longevity of service, and repair costs will determine the actual total cost of ownership over a pump's lifetime. When deciding on which materials for wetted pump parts, consider users' experiences, expected pump life, the hazards associated with the liquids being pumped, and the criticality of the service.

API 610 recommends materials of construction groups for a wide range of applications that are commonly found in petrochemical facilities. Recommended metallurgy for the casing, inner casing parts, shaft, and impeller are listed in API 610 materials of construction tables. For example, the API 610 S-6 (12% CHR) material class recommends the following materials be used as follows:

- Pressure casing: Carbon steel
- Inner case parts (bowls, diffusers, diaphragms): 12% chrome
- Impeller: 12% chrome
- Impeller wear rings: 12% chrome hardened
- Shaft: AISA 4140

Similarly, recommended materials of construction are listed for a wide range of pump applications in API 610.

The following criteria should be considered in the selection of the material for a centrifugal pump casing and its internal parts:

- Corrosion resistance
- Abrasive-wear resistance

- Cavitation resistance
- Strength (primarily for the casings)
- Casting and machining properties
- Cost

The Hydraulic Fit

To ensure a lifetime of reliable service from a pump, you must start with good hydraulic fit. A good hydraulic fit means that a pump operates close to its best efficiency point (BEP), which is where pump designers intended their pumps to operate. Operation away from the BEP results in unstable flow, pressure pulsations, vibration, and frequent pump failures associated with poor hydraulic fits. Unstable pump operation often leads to repetitive seal shaft and impeller failures.

%BEP at rated flow (Actual flow divided by the BEP flow) is a measure of how good a fit the pump is for your application. You want your %BEP of rated flow to be above 60% and hopefully over 80% and below 120%. In the pump curve in Figure 6.8, the best efficiency flow is at 1000 gpm, which means a design flow of less than 600 gpm would represent a %BEP flow of less than 60%. Similarly, a design flow of less than 800 gpm would

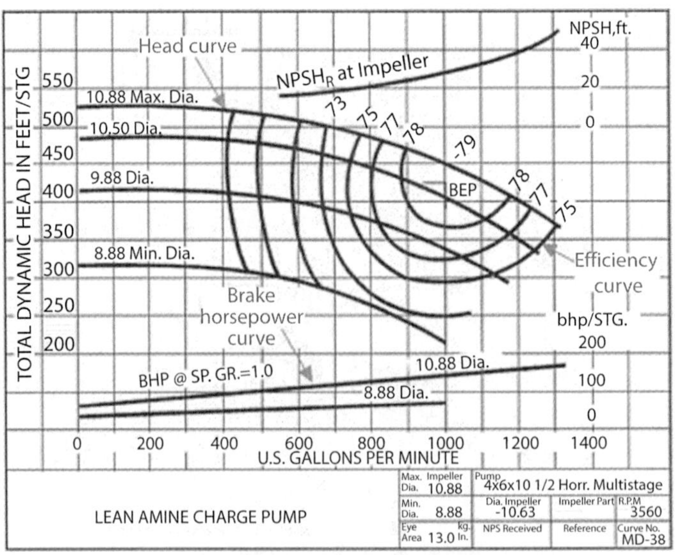

Figure 6.8 Typical pump curve. The BEP flow is 1000 gpm.

represent a %BEP flow of less than 80%. Remember that the BEP flow is the flow on the pump performance map where the efficiency peaks.

Not only does a pump need to operate close to its BEP flow but it needs to be able to operate reliably over the flow range required by the process at all possible operating conditions. For example, let's say a pump's BEP flow is 1000 gpm and that you expect the pump to operate between 400 to 800 gpm due to changing process demand conditions; this means that this pump is expected to operate anywhere from 40% to 80% of its BEP flow. The low end of the flow will present a problem, since we have stated that flow below 60% of the BEP flow is not recommended. There are two possible ways to handle this situation:

1. Install two 500 gpm rated pumps and run two pumps when you need to move 800 gpm and run one when you need to move 400 gpm. (In extremely critical pumping services, you would need to install three 500 gpm rated pumps. During times when you are operating two pumps, you would always have a spare in case one pump fails.)
2. Another possible solution might be to perform a hydraulic rerate (see description of hydraulic rerates below) on the pump. By rerating the pump internals for 700 gpm, the modified pump would be operating in the 57% of BEP to 114% of BEP flow range. One caveat with this solution is that you must ensure that there is ample NPSH available for the maximum flow condition.

Pump hydraulic rerates [3]: The basic idea of a pump hydraulic rerate is to fit an impeller with a lower flow rating inside the existing oversized pump casing. To accommodate a new impeller, case wear ring adapters are used to allow for smaller diameter sets of casing and impeller wear rings (see Figure 6.9). Rerate modifications also include matching the volute throat area(s) to the new lower flow impeller. Flow matching the impeller to the casing is accomplished by cutting out the original volute lips and welding in custom-designed volute lips.

This simple, straightforward design approach can provide pump users a cost-effective solution to chronic failures caused by flow conditions that are well below a pump's best efficiency point. Furthermore, if they make economic sense, power end design improvements, such as, a) upgrading to a larger shaft to improve the stiffness ratio (L^3/D^4), b) increasing the size of the stuffing box to improve seal reliability, c) upgrading the radial and

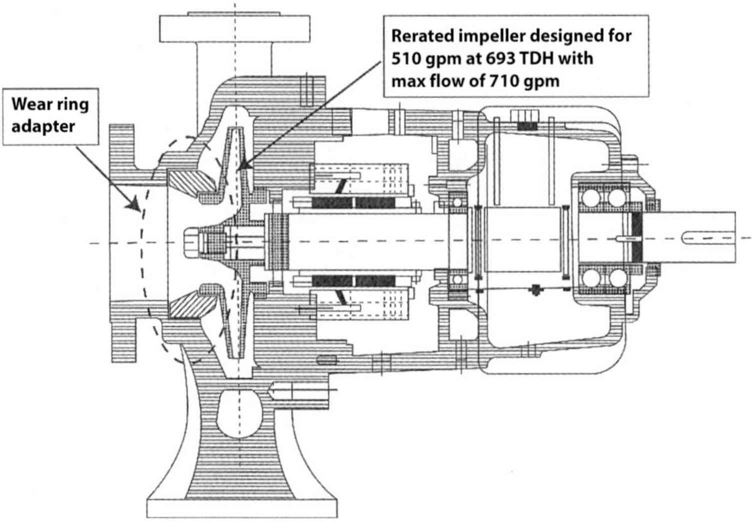

Figure 6.9 Cross section of a rerated single-stage centrifugal pump with new case wear ring adapter and power end upgrade.

thrust bearing, and d) adding bearing housing isolators, should be incorporated during the pump rerate.

The NPSH Margin [4]

To limit the amount of cavitation occurring in the first stage impeller, pumps need a margin of additional NPSH above the test stand values. It has been demonstrated that the amount of NPSH margin required for safe operation depends on the suction energy of the pump, which reflects energy available for cavitation damage. Suction energy is a function of the suction specific speed (N_{ss}) of the pump.

Table 6.5 NPSH margin ratio guidelines ($NPSH_A/NPSH_R$).

Suction energy level	
Suction Energy Level	Margin
Low	1.1-1.3
High	1.3-2.0
Very high	2.0-2.5

Table 6.5 lists the suggested NPSH margin ratio that should be maintained between the NPSHA and the NPSHR as recommended by the Hydraulic Institute.

In Table 6.5, margin is defined as the ratio of $NPSH_A$ to $NPSH_R$. The suction energy is a function of the momentum of the fluid approaching the impeller eye and is defined by the following equation:

Suction Energy = D_{eye} x n x N_{ss} x S.G.; where: D_{eye} = Impeller eye diameter (in) n = Pump rotational speed (rpm) N_{ss} = Suction specific speed (dimensionless)

$$N_{ss} = \frac{N \times \sqrt{Q_{BEP}}}{(NPSHr_{BEP})^{0.75}}$$

Q is flow in gpm

NPSHr is the net positive suction head required

S.G. = Specific gravity of the pumped fluid

<u>Threshold Values for End Suction Pumps are defined as:</u>

High suction energy 160 x 10^6

Very high suction energy 240 x 10^6

Let's go through a quick example using the guidelines in Table 6.5 to determine a safe $NPSH_r$ value: Assume you are considering an impeller with a 6" eye that will be operating at 3600 rpm, with a specific gravity of 0.8, and a N_{ss} of 10,500. The suction energy level in this application is:

$$SE=6 \text{ x } 3600 \text{ x } 10{,}500 \text{ x } 0.8 = 181.4 \text{ x}10^6$$

This SE value puts you in the "High" suction energy level range, which means that and NPSH ratio ($NPSH_a/NPSH_r$) between 1.3 and 2 is recommended. So, if the manufacturer states this impeller requires 20 ft of NPSH, then you should provide anywhere between 26 to 40 ft of $NPSH_a$ to ensure reliable operation.

Tip: When determining the NPSHa requirements for your specific application, make sure your calculation is based on the maximum expected flow. Since most manufacturers' NPSHr values rise as flow increases, you should always base your $NPSH_a/NPSH_r$ ratio calculation on the highest expected pump flow case.

Seal and Seal Flush Design

Mechanical seals can only operate reliably for years if they have been carefully selected for their specific application, installed properly, and operated with care. The seal designer must clearly understand the anticipated operating conditions so that the right type of seal and materials of construction are chosen to achieve maximum seal performance. Here are some key design review steps to follow if you want to be sure you have the right mechanical seal for the application at hand:

1. Identify the exact liquid being handled and its properties is the first step in the seal selection process. Seal face and hardware material must be able to withstand the fluid being processed. In other words, all seal materials must be chemically compatible with the fluid at the actual pumping temperature. In addition, the liquid's vapor pressure must be known to ensure the fluid being sealed remains in liquid as it flows across the sealing faces. It is also important to know the liquid's viscosity and solids content to ensure a reasonable seal life. Abrasive liquids can create accelerated wear that will ultimately shorten the seal's life. Double mechanical seals or an external seal flush plan can be used to cope with difficult fluids.
2. The pressure inside the seal chamber and the seal size determines the type of seal required and if it needs to be a balanced or unbalanced type seal. The pump manufacturer can estimate the expected pressure inside the seal cavity at normal pumping conditions; however, if frequent failures are occurring, it is recommended that actual stuffing box pressures be measured in the field. The seal operating pressure is a critical seal design parameter because it is used to determine if additional cooling is required to keep the liquid between the seal faces from flashing or if the seal face loading may be excessive for the face material combination being used.
3. Seal materials of construction must be carefully selected to handle the liquid at the anticipated operating temperature. It is vital to know a seal's maximum operating temperature because all seal material of construction will have upper temperature operating limits that should never be exceeded. Always provide the seal supplier with the full range of temperatures that the seal will experience. The design temperature range should even include a safety margin to cover process upset conditions.
4. Every plant has its own set of standards and operating procedures when it comes to mechanical seal reliability and

emission requirements. The seal type and arrangement selected must meet or exceed the desired reliability and emission standards for the pump application.

When selecting the seal's materials of construction, ensure the materials selected a) can be operated safely at the maximum expected operating temperature, and b) possess sufficient chemical resistance properties for the application. When evaluating seal face material combinations consider these material characteristics:

- Wear resistance
- Low leakage & friction properties
- Thermal expansion and conductivity properties of components and adaptive hardware
- Corrosion resistance

The API 682 is an excellent source for mechanical seal design knowledge and selection guidance. The goal of this standard is to assist users in the selection and operation of end face mechanical seals in centrifugal pumps used in the petroleum, natural gas, and chemical industries. In addition, annex G of this standard provides illustrations and a short tutorial about each piping plan. API 682 is based on the combined knowledge and experience of seal manufacturers, engineering companies, and end users.

Challenging Pump Applications

There are thousands of different pump applications found throughout the petrochemical industry. Most pump applications are normally trouble free and straightforward; however, there are some applications that present challenges to pump designers and users. We will discuss the following difficult applications found in petrochemical sites:

1. Pumps Operating in Parallel
2. Pumping liquids with low densities
3. Low NPSH services
4. Pumps Handling a Liquid with Varying Densities
5. Slurry Pumps
6. Bottoms Pumps
7. Hot Pumps with Galling Tendencies
8. High temperature applications

Pumps Operating in Parallel

Parallel pump operation is when two or more centrifugal pumps are installed side by side with common suction and discharge headers so that their flows are additive (Figure 6.10). Figure 6.11 is an example of two pumps that are installed in parallel, where both pumps are expected to share the pump load equally. Problems arise when operators assume the pumps are identical in head and flow characteristics and that flows will

Figure 6.10 Three centrifugal pumps installed in a parallel configuration. Either one, two, or three pumps can be operated to provide the desired flow rate.

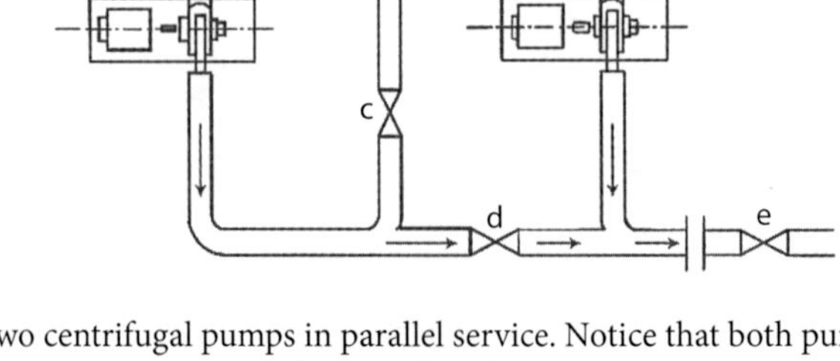

Figure 6.11 Two centrifugal pumps in parallel service. Notice that both pumps are connected to common suction and discharge headers.

Table 6.6 Parallel pump operating arrangements and safeguards. Note: The preferred operating range (POR) as defined by HI or API 610, varies depending on the type of impeller, but in most cases extends from 70% to 120% of flow at the best efficiency point (BEP). In some high-energy pumps, the POR is recommended to only extend from 80% to 115% of flow at BEP.

	Dissimilar pumps	**Identical pumps**
Constant Speed Operations	Special precautions are required for safe operation! 1. Ensure that all pumps are hydraulically matched so that they a) operate in the preferred operating range (POR), and b) generate adequate head to their lift check valves under all flow conditions 2. Provide some form of flow detection and/or protection for each pump: a) Individual flow meters or power meters for electric motors can be used to protect pumps from low flow conditions; b) low flow protection spillback valves may be required if operation below the recommended minimum flow guidelines could occur. 3. Butterfly valves may be needed on induvial pump discharge lines to balance pump flows as required. 4. Shut down pumps whenever possible to reduce energy consumption and start additional pumps when required by the process.	Viable arrangement with acceptable with safeguards 1. Ensure that all pumps are identical and that they a) operate in the preferred operating range (POR), and b) generate adequate head to lift check valve under all conditions 2. Provide some form of flow protection for each pump. Individual flow meters or power meters for electric should be used to protect pumps from low flow operation. 3. Since pumps never provide identical performance, butterfly valves may be used on pump discharge lines to help balance pump flows as required. 4. Shut down pumps whenever possible to reduce energy consumption and start additional pumps when required by the process.

(Continued)

Table 6.6 Parallel pump operating arrangements and safeguards. Note: The preferred operating range (POR) as defined by HI or API 610, varies depending on the type of impeller, but in most cases extends from 70% to 120% of flow at the best efficiency point (BEP). In some high-energy pumps, the POR is recommended to only extend from 80% to 115% of flow at BEP. (*Continued*)

	Dissimilar pumps	**Identical pumps**
Variable Speed Operations: with a variable speed motor, steam turbine, or gas turbine	Viable arrangement with acceptable with safeguards 1. Ensure that all pumps are hydraulically matched so that they a) operate in the preferred operating range (POR), and b) generate adequate head to lift their check valves over the entire speed range. 2. Provide some form of flow detection and/or protection for each pump; a) individual flow meters or power meters for electric motors should be used to protect pumps from low flow operation; b) low flow protection spillback valves may be required if operation below the recommended minimum flow guidelines could occur. 3. Provide a variable speed control arrangement that allows individual pump flows to be varied independently to ensure all pumps are operating in a safely throughout the expected flow range. 4. Employ a control strategy that minimizes the overall energy consumption by operating pumps close to their best efficiency point. Shut down pumps whenever possible to reduce energy consumption and start additional pumps when required by the process.	Ideal Arrangement 1. Ensure all pumps are hydraulically matched so that they a) operate in the preferred operating range (POR) and b) generate adequate head to lift check valve under all flow conditions 2. Provide some form of flow protection for each pump. Individual flow meters or power meters for electric motors should be used to protect pumps from low flow operation. 3. Provide a speed control arrangement that allows individual pump flows to be varied independently to ensure all pumps are operating safely throughout the expected flow range. 4. To obtain the best overall efficiency, identical parallel pumps should all be operated at the same flow conditions and as close to their best efficiency point as possible. Shut down pumps whenever possible to reduce energy consumption and start additional pumps when required by the process.

remain similar throughout their lifetime. However, if either an unexpected piping restriction develops, pump performance is not quite identical, or one of the pumps is more worn than the other, then significant flow differences between pumps can be observed. Safeguards for parallel pump operations may be needed to prevent or detect unsafe flow conditions.

Let's consider the four possible parallel arrangements seen in pumping applications. They are:

1. Dissimilar pumps operating at a constant speed
2. Identical pumps operating at a constant speed
3. Dissimilar pumps operating at variable speeds
4. Identical pumps operating at variable speeds

Each of these pumping arrangements presents the user with different levels of risk due to their likelihood of experiencing low flow conditions. For example, the "dissimilar pumps operating at a constant speed" represents the highest risk pumping arrangement, while the "identical pumps operating at variable speeds" represents the lowest risk pumping arrangement. Controls can be used to avoid unsafe flow conditions, while low flow detection methods can be used to prevent prolonged operation at unsafe flow conditions. Table 6.6 lists some ways to avoid unsafe low conditions when applying each type of pumping configurations.

Keep in mind that to ensure parallel pumping arrangements are safe and reliable, users must begin by selecting well-matched pumps. Different pump designs can be operated reliably in parallel service if they have similar shutoff heads and specific speeds. Next, good piping design practices must be used to ensure symmetrical flows with minimal turbulence are realized. Controls and safeguards will be ineffective in protecting parallel pump if the pumps are poorly selected and installed.

Pump Liquids with Low Densities

Centrifugal pumps can be designed to handle a wide range of liquid densities. However, the pump designer must keep in mind that as the pumpage gets lighter, the pump head required to generate a given differential pump pressure is inversely proportionate to the specific gravity of the liquids being pumped. The relationship between pressure and head is given below:

$$H = \frac{P \times 2.31}{S.G.}$$

Referring to Figure 6.12, we can visualize the relationship between head and pressure. For example, a pump that can generate 50 psi pumping a liquid with a specific gravity of 1.0 can only generate 35 psi pumping a liquid with a specific gravity of 0.70. In general, we can say that the pressure generated for a fixed pump head decreases proportionally with the density of the liquid being pumped.

To select the right pump, we must first determine the specific speed required for the application. Specific speed (N_s) is defined as "the speed of an ideal pump geometrically similar to the actual pump, which when running at this speed will raise a unit of volume, in a unit of time through a unit of head". N_s, which defined below, is an index used to predict desired pump performance and the general shape of a pump's impeller.

$$N_s = \frac{N \times \sqrt{Q_{BEP}}}{(H_{BEP})^{0.75}}$$

Where (in U.S. Units):

N_s = specific speed
N = rotational speed of impeller
Q_{BEP} = capacity of pump at the best efficiency point
H_{BEP} = head of a single stage of the pump at the best efficiency point

Once the desired specific speed is known, basic dimensions of the unit's components can be easily calculated. Centrifugal pumps are traditionally

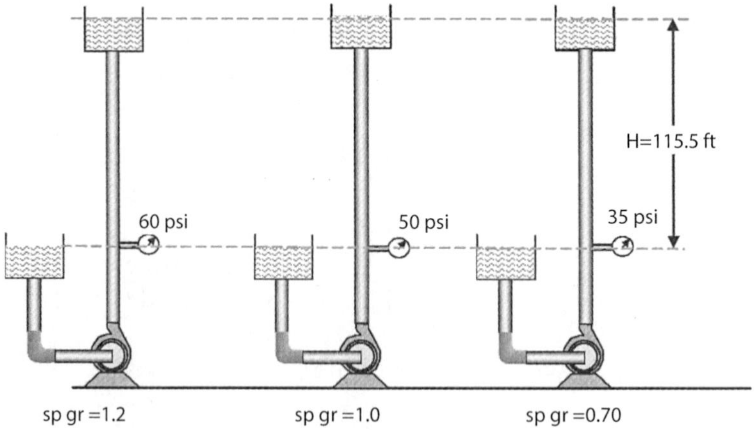

Figure 6.12 A constant speed, fixed head pump will generate more differential pressure when moving heavier liquids than when moving a lighter liquid.

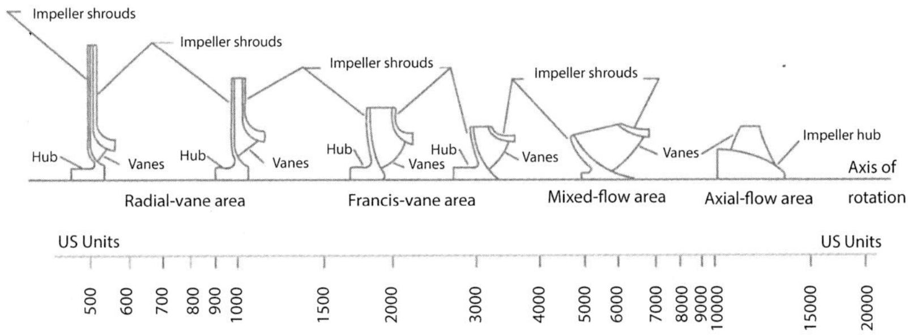

Figure 6.13 A comparison of centrifugal pump impellers with different specific speeds. As the specific speed increases, the flow through an impeller transitions from a fully radial flow (Ns=500) to a fully axial flow (Ns>10,000).

divided into three types: radial flow, mixed flow, and axial flow. Notice in Figure 6.13 there is a gradual change from the radial flow impeller, which develops pressure principally by the action of centrifugal force, to the axial flow impeller, which develops most of its head by the propelling or lifting action of the vanes on the liquid.

Using the relationship between pressure and head (H = P x 2.31/S.G.) that we discussed earlier, let's look at two pumping cases. First, let's see how much head an impeller with a specific speed of 1000, turning at 3600 rpm and moving 500 gpm can develop if it's moving water (S.G. = 1). We see from the calculation that this impeller will generate 347.6 feet of head or 144 psi (349.6/2.31) of differential pressure.

First, let's see how much head an impeller with a specific speed of 1000, turning at 3600 rpm and moving 500 gpm can develop if it's moving water (S.G. = 1). We see from the calculation (Table 6.7) that this impeller will generate 347.6 feet of head or 144 psi (347.6/2.31) of differential pressure. If this same impeller (Ns = 1000) were pumping propane (Specific Gravity (at 60° F) = 0.504), the same 347.6 feet of head would equate to only 75.8 psi of differential pressure. In other words, assuming identical impellers, it takes approximately twice as many of impellers pumping propane, to generate a given differential pressure than if we were pumping water. You have probably noticed in the field that vertical turbine pumps that handle lighter liquids typically have more stages than those pumping water.

Table 6.7 Example pump head calculation for a hypothetical Ns = 1000 impeller.

Symbol	Variable	Value	Units
Ns	Specific Speed	1000	specific speed
N	RPM	3600	rpm
Q	Flow Rate	500	gallons/minute
H	Head	347.58	feet

The density of water hardly varies with temperature; however, density of an LNG mixture can change significantly with slight changes in its composition. This sensitivity of liquid density to composition explains why when working with lighter liquids it is so important to know the exact composition of the pumpage. Slight variations in product composition can change a pump's performance significantly and result in field problems. The worst situation is when the liquid's density gets so low that the pump cannot develop sufficient pressure required to move liquid downstream.

Low NPSH Services

Consider the reflux pump in Figure 6.14. We have seen that liquids at the top of a fractionator tend to be lighter and more volatile. In the configuration seen here, a portion of the vapor coming off the top of the tower is condensed and drops into an overhead receiver. Some receivers will have a boot to collect liquids so they can be pumped out. Depending on the efficiency of the condenser, the liquid that collects inside the receiver will have a temperature slightly below its bubble point. (The bubble point of a mixture is the temperature—at a given pressure—where the first bubble of vapor is formed when heating a liquid consisting of two or more components.) If we assume the liquid inside the receiver is at its bubble point, which is a common design assumption, then the net positive suction head available is simply the head of liquids above the pump's suction nozzle. So, if the level inside the receiver is 20 feet above the suction of the pump's suction nozzle, then we assume we have 20 feet of NPSHa.

In applications where the NPSH margin is known to be small, designers may need to either raise the level of the receiver or resort to vertical turbine pumps as seen in Figure 6.15. The author has resorted to using vertical can pumps in applications where horizontal pump options were not viable.

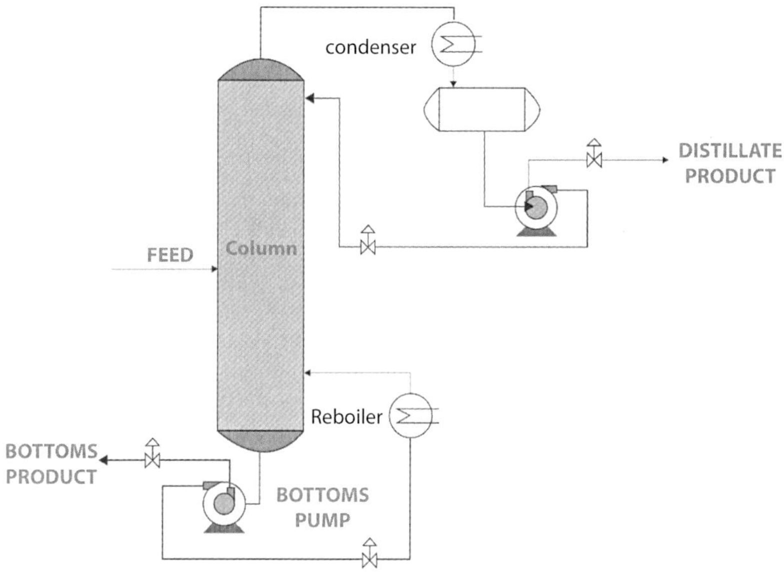

Figure 6.14 Typical reflux pump arrangement at the top of a fractionator. Vapors off the top of the tower are cooled with a heat exchanger. Liquid that condenses in the overhead receiver can either be pumped back into the fractionator or pumped to another location.

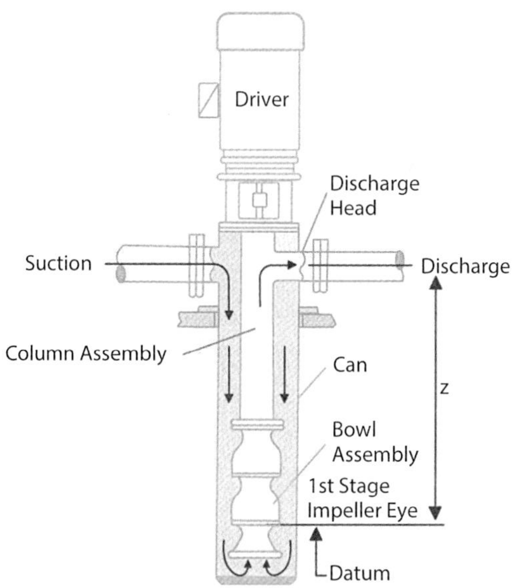

Figure 6.15 Vertical turbine centrifugal pump. The distance from the centerline of the suction nozzle to the eye of the 1st stage impeller (Z) is considered to be additional NPSH available in the overall NPSH available calculation.

Vertical turbine centrifugal pumps by their design can add elevation head to the suction conditions and increase the available NPSHa.

How an Impeller's Suction Specific Speed Affects the Required NPSH [4]

Let's examine how an impeller's geometry and speed can affect the minimum required NPSH. Figure 6.16 summarizes the minimum net positive suction head required (NPSHr) for a range of flow and rotational speed conditions. If you discover that the impeller you are considering has a minimum required NPSH that is greater than the available NPSH, then you need to either a) increase your NPSHa, b) consider a different impeller, or c) change your design speed.

Notice that Figure 6.16 contains four different curves: An NPSHr versus flow line for an 1800 rpm, single suction pump, 3600 rpm, single suction pump, an 1800 rpm, double suction pump, and a 3600 rpm, double suction pump. Here a quick example of how to use this chart: Let's assume you want to pump 6000 gpm with an impeller with single suction eye, operating at 3600 rpm. Based on Figure 6.16, this impeller will require a minimum of

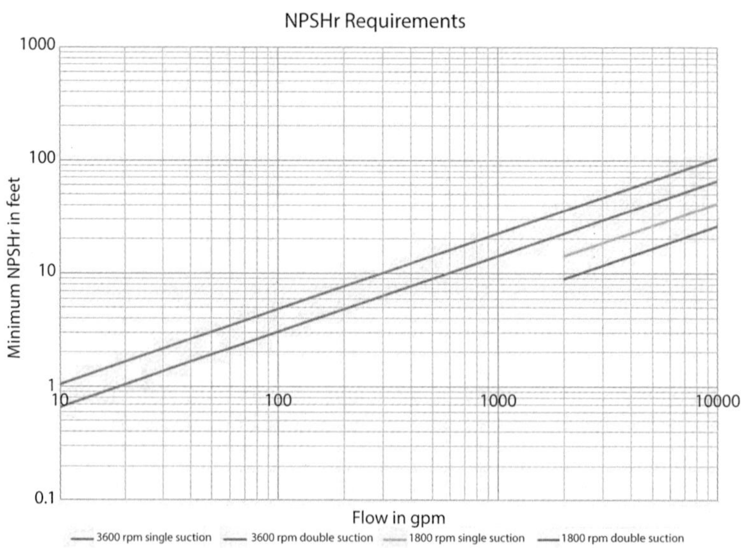

Figure 6.16 NPSHr versus flow for various impeller designs in log-log format. (Note that this NPSHr chart is based on an accepted design practice of keeping the suction specific speed below a recommended value of below 11,000. Impellers with values higher than 11,000 will have even higher NPSHr requirements.)

about 75 feet of NPSHr. Dropping the speed down to 1800 rpm will allow you to drop the NPSH requirement down to about 28 feet with a single suction impeller.

Figure 6.16 can quickly show you the minimum NPSH that will be required depending on the speed and impeller geometry you are considering. However, be careful when trying to apply the curves in Figure 6.16. These curves assume flow at the inlet impeller's best efficiency point and no NPSH margin. Figure 6.16 is meant to serve as a starting point in the pump selection process. Always start by using the manufacturer's actual $NPSH_r$ curve and select the $NPSH_r$ for the maximum expected flow; then determine a healthy NPSH margin based on the suction energy (see "NPSH margin" section) for the application. Finally, use the selected NPSH ratio to determine the required $NPSH_a$.

The primary lesson to be learned from the Figure 6.16 is that a lower impeller NPSHr value can be obtained by either lowering the design pump speed or doubling the impeller eye area by using a double suction impeller. However, never blindly apply impellers with higher suction specific speed or double suctions just to lower the NPSH requirements. Impellers with specific speeds over 11,000 and those with double suction impellers are prone to flow instability when operated away from their best efficiency point. From a reliability point of view, it's always better to keep your suction specific speeds below 11,000, use double suction impellers judiciously, and reduce the pump's design speeds to the lowest speed practical speed.

When dealing with low NPSHa applications:

1. Make sure you design for lowest liquid level situation. Many processes designate "low level" and "low-low level" alarms for liquids levels. If possible, design your pump for "low-low level" situations.
2. Provide a healthy margin between the NPSHa and NPSHr. Add additional NPSH margin for pure liquids.
3. Always consider high flow cases to determine NPSH margins.
4. Ensure there is sufficient retention time on the overhead receiver to properly disengage vapors. Adding a boot to increase the settling volume or installing settling weirs can help promote vapor disengagement.

Here are some suction piping guidelines for pumps handling liquids with small NPSH margins:

- o Keep suction piping as short as possible.
- o Be generous with suction piping diameters to minimize head losses. The suction piping diameter should be equal to or greater than the suction nozzle diameter.
- o Minimize the use of elbows to reduce head losses.
- o Field experience has shown that the ideal minimum length of full-size, straight pipe immediately upstream of the suction nozzle is 10D (10 pipe diameters) and the absolute minimum length of full-size, straight pipe should be 5D.
- o Continuously slope suction piping from the suction vessel all the way to the pump suction nozzle.
- o Never allow high points in the suction piping that would allow vapors to collect and possibly vapor lock the pump.
- o If a reducer is required in the suction lines, install it with the flat side to the top.

Pumps Handling a Liquid with Varying Densities

Pumps that handle light liquids that vary in density can present a design challenge. Hot wells or overhead vessels that handle mixture of light liquids can be especially tricky. As the mixture gets lighter, the ability of the pump to generate differential pressure is reduced. In extreme cases, the pump will not be able to generate the discharge pressure to lift the discharge check valve and establish flow.

One application that was particularly difficult was a liquid collection drum for a flare header. Liquids that were collected had a wide range of densities. There may be occasions when the head developed by the pump was unable to generate the pressure to open the check valve.

There are several ways to design pumps to handle wide variations in liquid densities:

- Use a variable speed pump driver to enable the pump to generate the required head under all conditions and stay within a healthy flow range.
- Increase the head capability of the pump to generate sufficient pressure and lower densities. This is a less energy efficient solution, because when liquid density is high, the flow control valve will have to drop the excess pressure down as required for the discharge system.

Let's go through a simple example to illustrate how wasteful it is to use a constant speed pump in a service that is expected to handle liquid expected to vary widely in density. Assume we wish to select a constant speed pump capable of moving 250 gpm from 20 psig into 120 psig system. If the liquid density of the mixture varies from 0.67 to 0.87, the designer needs to pick a pump capable of generating a 100 psi + 20 psi (for control) psi of differential pressure for the full range of liquids. When pumping 0.67 S.G. liquids the pump must be rated for 414 ft of head; however, when pumping the 0.87 S.G. liquids, the pump will deliver 120 x (0.87/0.67) = 155.8 psi of differential pressure. The difference between the pressure produced by the pump moving with the heavier liquid and pressure produced by the pump moving the lighter liquids will have to be removed by the flow control valve.

The additional 35.8 psi pressure drop across the control valve is equivalent to about 8 brake horsepower (34.98 HP-26.94 HP), which could be avoided by using a variable speed pump driver. Installing a variable speed driver would reduce the horsepower draw by 30% by allowing the pump to drop in speed when moving heavier liquids. Another advantage of installing a variable speed driver would be its ability to provide overspeed capability in case more head or flow is required for upset conditions.

- Install a pump flow alarm to alert operators that the pump is running but is not moving liquids forward. Another option might be to have an alarm whenever there is a drum high level, which would indicate that liquid is not being pumped out of the drum fast enough to keep up with the inflow.

Slurry Pumps [5]

Anyone who has had to pump a slurry knows that it can be one of the most challenging fluid applications. The more you understand about what you're pumping, the better your pump selection becomes, allowing you to achieve longer mean time between failure. Slurries are abrasive, thick, sometimes corrosive, and contain high amounts of solids. Some aggressive slurries can significantly reduce the life of a pump due to erosion at high velocity areas.

What is a Slurry?

Slurry is any mixture of fluid and fine solid particles. Slurries are used as a convenient way to handle solids in bulk in mining, steel processing, foundries, power generation, and most recently, the Frac Sand mining industry. In some cases, such as in cat cracker slurry pumps, catalyst fines are carried along with the process liquids. Slurries generally behave the same way as thick, viscous fluids. Furthermore, slurries can be highly erosive, depending on the size and geometry of the suspended particles.

Slurries are divided into two general categories: non-settling or settling. Non-settling slurries consist of fine particles, which give the illusion of increased apparent viscosity. These slurries usually have low wearing properties but do require careful consideration when selecting the right pump because they do not behave in the same manner as a normal liquid does. Settling slurries are formed by coarse particles that tend to form an unstable mixture. Particular attention should be given to flow and power calculations when selecting a pump. Many slurry applications are made up of coarse particles and because of this, have higher wear properties. Here is what you should know about your slurry application:

- Abrasiveness
- Consistency
- Percentage of solids by weight
- Settling time
- Power correction requirement because of density of the slurry

Basic pump components, such as the impeller size and geometry, material of construction, and discharge configurations must be considered to ensure the pump will hold up against the wear caused by an abrasive slurry. Slurry pumps are generally larger in size when compared to low-viscosity liquid pumps and usually require more horsepower to operate because they're less efficient. Bearings and shafts must be more rugged and rigid as well.

In general, slurries tend to be heavy and difficult to pump. Not only do slurries cause excessive wear on pumps, but they are also known to clog suction and discharge lines if the slurry flow is not moving fast enough. Most importantly, it's a challenge to make slurry pumps last for a reasonable amount of time. But there are a few things you can do to extend the life of your slurry pump and make pumping slurry less of a challenge:

To reduce wear, operate as close to the BEP as possible. Any significant flow turbulence created at off design conditions will drastically increase the erosion rates.

Reduce the pump's discharge pressure as much as possible to allow operation at the slowest possible speed, but fast enough to keep solids from settling and clogging the lines.

Follow good piping practices to ensure a constant and uniform flow of the slurry into the pump.

FCC Slurry Pumps

One especially difficult slurry application found in Fluidic Cat Cracker (FCC) units are the fractionator bottoms pumps. These pumps recycle the slurry oil from the fractionator back to the reactor feed system. The slurry oil consists of a mixture of the heavy oil in the bottom of the fractionator and suspended catalyst particles that carry-over from the reactor into the fractionator. The slurry oil pumps are specially designed to withstand the erosive action of the catalyst powder suspended in the pumpage. These pumps usually handle oil flows in the range of 500 to 1000 gpm, at 600 to 700°F. Usually, they vertically split overhung or between bearing centrifugal pumps.

Since the pressure differential between the fractionator and reactor is low, the heads on these pumps are low. They are designed to operate at reduced speeds (1180 to 1750 RPM range) to reduce erosion of the catalyst in the slurry. In addition, clean flushing oil is supplied to the wear rings, throat bushings and packing, through specially designed connections, to reduce erosion in wearing parts. Because of the temperatures and sulfide environment, the pumps are typically furnished with chrome cases and impellers.

Bottoms Pumps

Pumps located at the bottom of refinery towers tend to handle more solids and trash due to the tendency for gravity to force free particles or heavy fragments downward. Broken tray parts, solids in the tower feed, coke particles, etc., always seem to find their way to the bottom of towers and into "bottoms pumps." Operational upsets can occasionally damage tower trays or result in high temperatures that can generate excessive coke particles. It should be expected that operational upsets will result in more unexpected material passing though bottoms pumps and into the pump discharge piping.

If utilized, a bottoms pumps suction strainer may be sometimes found to be plugged up with tray parts, coke, etc., after upsets. There are several ways to design for high solid loads in bottoms pumps: a) ensure the impeller (or impellers) can pass particle sizes that you are expecting to see, or b) install suction strainer with opening that are smaller than the maximum particle size that the impeller can handle. Whenever a bottoms pump experiences a loss in pressure or flow, always consider some type of suction pluggage first. A differential pressure transducer may be used to detect strainer plugging before it becomes a problem.

Another concern related to bottoms pumps can be having to handle high viscosity liquids whenever changes in the process occur. We have mentioned the density gradient created inside of a distillation tower. As we move down processing towers, the density and viscosity of the liquids increase. The liquids that end up at the bottom of the tower are made up of a mixture of heaviest molecules. The bottoms found in Crude Vacuum towers and Coker fractionators can have high densities and viscosities compared to the other crude fraction. If the viscosity rises significantly due to a drop in operating temperature or a change in composition, pump performance will be affected. Since liquid density and viscosity can affect pump performance, we need to know the actual properties of the liquid being pump. But before evaluating pump performance, you should check to see the unit operating conditions are normal.

Figure 6.17 illustrates how viscosity affects centrifugal pump performance. The solid lines represent pump performance without viscosity effects and the dotted lines represent pump performance due to a 1000

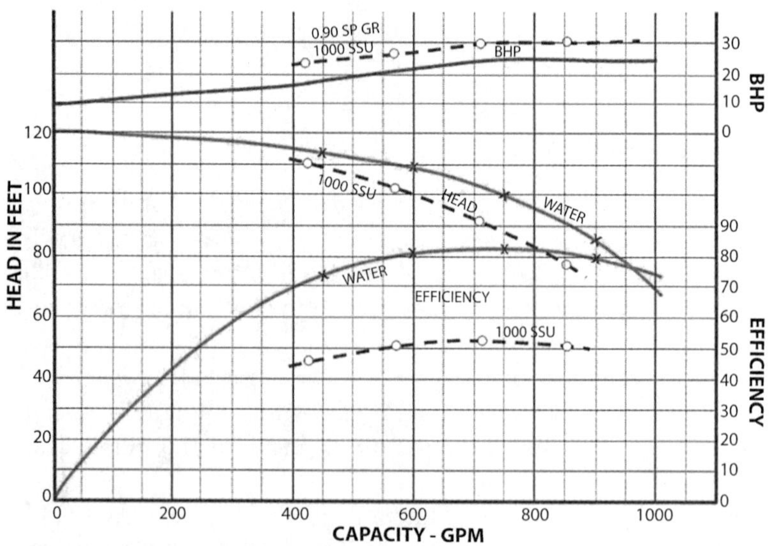

Figure 6.17 Effect of viscosity on centrifugal pump performance.

Table 6.8 Comparison of expected centrifugal pump efficiency at various viscosities. Assumptions: SG = 0.9 and Baseline Eff = 75%.

Viscosity (Cst)	**100 gpm @150 ft: expected efficiency**	**100 gpm @150 ft: efficiency loss**	**250 gpm @ 200 ft: expected efficiency**	**250 gpm @ 200 ft: efficiency loss**	**1000 gpm @ 325 ft: expected efficiency**	**1000 gpm @ 325 ft: efficiency loss**
10	70.0%	6.7%	69.0%	8.0%	72.0%	4.0%
50	56.0%	25.3%	59.0%	21.3%	65.0%	13.3%
100	46.0%	38.7%	52.0%	30.7%	60.0%	20.0%
250	29.0%	61.3%	40.0%	46.7%	51.0%	32.0%
500	15.0%	80.0%	30.0%	60.0%	43.0%	42.7%
1000	7.0%	90.7%	17.0%	77.3%	32.0%	57.3%
2000	4.0%	94.7%	7.0%	90.7%	18.0%	76.0%
3000	0.0%	100.0%	8.0%	89.3%	11.0%	85.3%

SSU fluid. Notice how the developed head falls below the theoretical head curve and that efficiencies fall far below those expected for inviscid fluids. As a consequence of the lower expected efficiencies, the predicted power levels rise above the manufacturer's standard horsepower curve.

Table 6.8 is a summary of the effects of viscosity on three different centrifugal pumps. Each pump is represented with two columns (expected efficiency and efficiency loss due to viscosity). As an example, let's look at a 250 gpm, 200-foot pump that is pumping a 100 cSt viscosity liquid. Under these conditions, the efficiency drops from 75% to 52%. Now let's see what happens when this same pump must pump a 1000 cSt liquid. The efficiency drops from 75% to 17%. We can see several trends by studying this table: First, you see that efficiencies drop off and viscosities increase. You also notice the performance of smaller pumps tend to be affected more significantly by higher viscosity liquids. Finally, you can see that there is a point when efficiencies start to drop off dramatically.

As we have shown here, higher liquid viscosities can significantly affect pump performance, which may require pumps and drivers to be oversized for the required performance. Be sure to size pumps and drivers for high viscosity excursions if they are likely to occur. There is a point when a centrifugal pump may not make economic sense. Carefully evaluate the total life cycle cost when considering using a centrifugal to pump a viscous liquid.

> Tip: While many centrifugal pump manufacturers say they can handle fluids with viscosities 1,000 cSt and higher, positive displacement pumps tend to be a better choice above these viscosity levels when considering the high energy costs resulting from the loss of efficiency.

Hot Pumps with Galling Tendencies

Some impeller wear rings tend to gall when the rotating and stationary rings are made from the same material. Galling is a form of wear caused by adhesion between sliding surfaces, whenever contacting surface are forced together. Galling accelerates whenever two metals are in close running clearance and develop a burr either due to abrasives in the pump media or rubbing. Stainless steels are examples of wear ring metallurgy that are particularly susceptible to galling. Typically, wear rings are designed with a hardness difference between the two rings to prevent galling (refer to Table 6.4).

Table 6.9 Recommended wear ring clearances.

Wear ring diameter	Diametrical clearance under 500 °F	Diametrical clearance under 500 °F
3.5"- 5"	0.016"	0.018"
5" – 6"	0.017"	0.019"
6" – 7"	0.018"	0.020"
7" – 8"	0.019"	0.021"
8" – 9"	0.020"	0.022"
9" – 10"	0.021"	0.023"
10" – 11"	0.022"	0.024"
11" and over	0.023"	0.025"

When using wear ring metallurgical combinations that are susceptible to galling, add additional clearance to prevent contact between wear rings and ensure adequate preheating to ensure proper internal clearances are present. For pumping temperatures of 500°F and over, add 0.010" to the wear ring clearance (see Table 6.9). Also, whenever galling-prone wear ring materials such as stainless steel are used, 0.005" should be added to the clearance.

Starting Hot Pumps [6]

It is good practice to preheat hot service pumps before starting them. A major concern for pumps using wear rings with galling tendencies is that of rubbing between the rotating and stationary wear rings. Rubbing can be caused by either pump casing distortion or differential expansion between the rotor and casing. Users don't have any control over casing distortion or poor piping design, but they can control the differential growth between the rotor and stator by preheating their hot pumps.

One common means of preheating a pump in hot applications is by installing a small bypass (1" or smaller as required) around the pump's discharge and check valve (see Figure 6.18). For larger pumps, a better practice is to install warm-up piping that allows hot liquid to flow into the bottom of the casing and out the discharge nozzle to ensure the entire pump is adequately preheated. Multiple warm-up lines are recommended

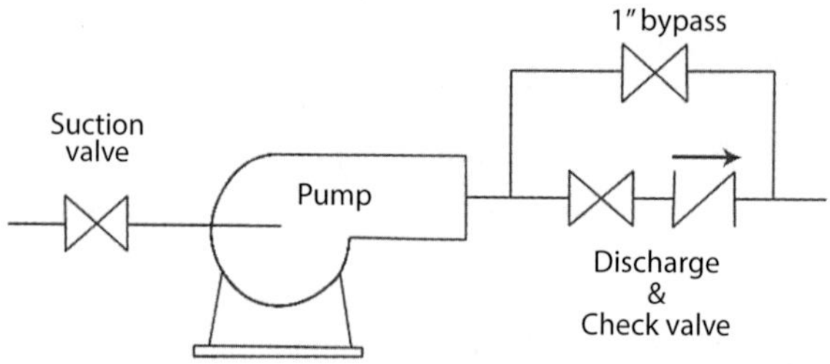

Figure 6.18 A hot pump with warm-up piping that allows liquid to bypass the discharge valve and check valve when it is not in service. Hot liquid flows from the discharge header back, through the bypass line and pump, and then back to the suction header. Note: A ¾" bypass valve or a ¾" valve with a flow restriction orifice can be used to control spillback flows on smaller centrifugal pump installations.

at the bottom of large multistage pump casings to evenly distribute internal flow to ensure thorough pre-heating. Do not forget to safeguard the mechanical seal's temperature environment. To prevent damaging the seal, mere preheating of the pump casing prior to startup may not be sufficient; protection of seal components against twisting in semi-solid liquid environments may be needed. Therefore, remember to preheat the seal region as well to prevent seal damage at startup.

Any pump with galling tendencies that operates above 150°F should be preheated prior to start-up. The risk of galling or grabbing is proportional to the pump's operating temperature. So, the hotter the pump operates the more care should be taken to ensure it is properly preheated before putting it in service. Insulation should be used to assist in the warm-up process and to maximize the thermal growth of the casing. Prior to start-up an infrared temperature gun or contact thermometer can be used to ascertain the casing is properly heated.

High Temperature Concerns

As we have mentioned, bottoms pumps tend to run higher temperatures, which leads to a concern over their materials of constructions. Designers must ensure flange gaskets and mechanical seal o-rings are capable of working reliably at pumping temperatures.

Gaskets [7]

Flange gaskets are used to create a static seal between two flange faces, at various operating conditions, with varied pressure and temperature ratings.

Table 6.10 The following table lists the pressure-temperature limits for various gasket types based on the process service, piping temperature and pressure rating, and flange facing design. Note: Some non-hydrocarbon pump services, such as hydrogen, steam, boiler feedwater, etc., are included in this table for completeness.

Service	Flange design conditions			
	Pressure class	**Temp °C**	**Flange rating**	**Gasket selection**
General Hydrocarbon	150 300	-196 to +500 (a) -196 to +500 (b) -196 to +350 (c)	RF	a. Tanged Graphite Sheet or b. Spiral Wound with Flexible Graphite or c. Spiral Wound with Non Graphite Filler
Steam/ Condensate, Boiler Feed Water				
General Utilities		-40 to 250	RF	Nitrile Rubber Based Reinforced Sheet
General Hydrocarbon, Steam/ Condensate, Boiler Feed Water	600 900	-196 to +500	RF	Spiral Wound with Flexible Graphite
General Hydrocarbon, Steam, Boil Feed Water	1500 2500	Based on materials of construction	RTJ	Metal Joint Ring

(*Continued*)

Table 6.10 The following table lists the pressure-temperature limits for various gasket types based on the process service, piping temperature and pressure rating, and flange facing design. Note: Some non-hydrocarbon pump services, such as hydrogen, steam, boiler feedwater, etc., are included in this table for completeness.

	Flange design conditions			
Service	**Pressure class**	**Temp °C**	**Flange rating**	**Gasket selection**
Hydrogen	150 300 600	-196 to +500	RF	Spiral Wound with Flexible Graphite
	900 1500 2500	Based on materials of construction	RTJ	Metal Joint Ring
Chemical Oxidizers/HF Acid	150	-40 to +200	RF	PTFE (reinforced or enveloped)
	150 300 600	-40 to +200	RF	Spiral Wound with PTFE Filler

A gasket fills the microscopic spaces and irregularities of the flange faces, which forms a seal that is designed to keep liquids and gases from leaking to the atmosphere. Materials for gaskets can typically be divided into three main categories: non-metallic types, semi-metallic types, and metallic types. There are various considerations when selecting a gasket which include fugitive emissions requirements, compatibility with the fluid, and operating temperature and pressure. The gasket selected should be suitable for the service and have reasonable life expectancy at the maximum operating temperature. Other special conditions might include vibration, erosive media, risk of the gasket contaminating the medium, corrosion of flanges, integrity, and the need for economy. Table 6.10 includes recommendations for flange gaskets in common piping applications.

Gasket materials are designed to compress under load to achieve the initial seal. However, to retain that seal, the gasket should be able to resist flow (or creep) to prevent loss of surface stress by bolt reduction.

Resistance to flow is an important gasket property and is the one that most readily separates high-quality from low-quality gaskets. Under ambient temperature conditions, most gasket materials do not creep significantly, but as the temperature rises beyond 100°C, creep becomes a serious consideration.

O-Rings

Most mechanical seals employ o-rings, which are composed of elasometic materials, for secondary sealing. They represent the seal's weakest link, which is easily affected by high temperature and chemical attack. For this reason, it is vital to know the temperature limits of your seal's elastomers and if they are compatible with the fluid being sealed. Table 6.11 lists common elastomers along with their temperature limit.

You should always look for signs of melting, extrusion, or disintegration on your failed o-rings once the seal is removed and decontaminated. If you see signs of o-ring disintegration, you should suspect elastomeric incompatibility. There are many o-ring chemical compatibility tables that are available for your reference online. If you see melting or extrusion, the operating temperatures are probably too high for your o-rings. Be aware that process excursions can result in operation outside the intended design range of pump o-rings and mechanical seals.

Table 6.11 Temperature limits of common elastomers.

Elastomer	Temperature range °F	Temperature range °C
Flourocarbon (Viton)	-15 to 400	-25 to 205
Ethyle propylene	-70 to 300	-55 to 150
Chemraz	-20 to 450	-30 to 230
Kalrez	0 to 500	-20 to 260
Neoprene	-45 to 300	-45 to 150
Buna N	-65 to 225	-55 to 105
Buna S	-75 to 250	-60 to 120

How Processing Issues Can Affect Pump Reliability

Process issues can result from either a) intentionally operating a process outside of its normal conditions, b) from an unexpected equipment or instruments failure, or c) unexpected change in the composition of the feedstock.

These issues, which may be due to operating errors and unforeseen events, should be expected to occur from time to time. Production problems can lead to flow and temperature excursions and some upsets can result in density changes due to changes in the liquid composition or vapor entrainment. In the worst cases, these production excursions have the potential to damage equipment.

There are many reasons that a process can be upset, but let's examine a few common causes:

- *Operating outside of processing design guidelines:* This situation occurs whenever operations are forced to run at higher or lower flowrates than the process was design for. Operators usually try to stay within the operating limits of their processes but sometimes operating outside pump preferred flow ranges cannot be avoided.
- Centrifugal pumps don't like to operate too far away from their best efficiency point. Operating a unit at an abnormal flowrate forces pumps to operate at abnormal flows. If a pump is found to be vibrating, noisy, or the flow is unstable, check with the operators to see if the unit is operating within its preferred flow range. The worst situation would be to run a pump below its safe minimum flowrate.
- *An unexpected pump or compressor failure:* An unexpected and sudden rotating machinery failure can result in a domino causing subsequent flow upsets upstream and downstream of the failure. Operating procedures should be written and followed to handle unexpected machine failures.
- *A control system failure:* An instrument failure or calibration issue can force pumps to operate incorrectly. Operating a pump at the wrong flow or speed can be problematic. Always check to make sure the controls are working properly before blaming the problem on a pump issue.
- *Partially plugged or fouled heat exchanger:* A plugged, or fouled heat exchanger can mean liquids will run hotter

or cooler than expected depending on the service, which can affect fractionation efficiency. In some cases, running too cool can affect pump performance. Running too hot could cause mechanical seals to fail due to product flashing across the seal faces or exceeding the material temperature limits.

- *Plugged strainers:* A plugged strainer is a common occurrence in a petrochemical facility. Plugging on the suction side of a pump can result in a loss of flow and in severe cases cavitation. A plugged strainer on the discharge of a pump will result in a drop in flow due to an abnormally high back pressure.
- *Operating with too low a tower or receiver level:* Low levels can result in a loss of suction head, resulting in cavitation. A low tower or receiver level can also lead to vapor entrainment due to a lack of retention time. Make sure there is ample retention volume to allow time to disengage vapors from the liquid in the vessel before entering the suction piping.
- *Using a feed that is out of spec or contaminated:*

    1) Running a feedstock that is out of the normal range of composition can have an immense effect on intermediate product flowrates and fluid properties, which in turn could affect pump performance. As suggested before, if a pump is found to be vibrating, noisy, or the flow is unstable, check with the operators to see if the unit is operating under normal conditions.
    2) There are too many types of contamination to discuss here. However, we will briefly note that the worst contaminant that can be found in a crude feed stream is water. A sufficient volume of water in the crude tower feed can lead to a severe flashing, which can damage tower internals.

No pump design can be expected to handle every type of process upset. However, there are ways to help most of your pump survive upsets. First, make sure you select the right design standard, i.e., the API 610 standard for severe application. Next, ensure your pumps have a) generous NPSHa margins to handle level upsets, b) are specified with excess horsepower to handle changes in liquids properties, and c) have N_{ss} values below 11,000 to provide a wider allowable stable flow range. Finally, provide operators with clear operating limits for your pumps and ensure there are operating procedures for start-ups, normal operating conditions, upset conditions, and shutdowns.

Summary

We have seen that petrochemical processes produce a range of liquids with different liquid densities, viscosities, and temperatures. The properties of the feedstock and how it is processed largely determines the physical properties of the subsequent intermediate hydrocarbon products that must be handled. While every pump application is unique, there are some applications that challenge even veteran pump designers. Here is a quick review of what we have learned about difficult pump applications:

1. When running pumps in parallel, provide adequate controls to keep all pumps operating in their preferred flow ranges. Additionally, low flow alarms and trips and/or minimum flow spillbacks may be used to protect parallel pumps for operating at unsafe flows for prolonged periods of time.
2. Lighter liquids, i.e., liquids with specific gravities lower than 0.6, require more pump head to develop the same differential pressure as those pumping heavier liquids; so, pumps in these applications typically required more pumping stages.
3. In applications where liquids can vary in density due to the nature of the process, a variable speed driver may be required to enable the pump to generate the required head under all conditions. Tip: Select a pump capable of delivering the required differential pressure when pumping the lightest expected liquid.
4. In applications handling light, volatile liquids, ensure there is an ample NPSH margin.
5. For pumps handling high-temperature liquids, design for the maximum possible operating temperature (MOT). Ensure gaskets and o-rings are designed for the expected MOT.
6. In slurry applications: a) to reduce wear, operate as close to the BEP as possible. Eddies created at off design conditions drastically increase erosion; b) reduce the pump's discharge pressure as much as possible to allow operation at the

slowest speed, but fast enough to keep solids from settling and clogging the lines.

7. Bottoms pumps tend to have trash and solids in pumpage. If sizable trash is expected, install suction strainers with sufficient trash capacity to handle the volume of trash that is expected.
8. Be careful when starting high temperature pumps with galling tendencies. Preheat all pump prone to galling prior to start-up that are designed to operate above 150 °F.

Once users clearly understand a pump's operating conditions and the physical properties of the product being pumped, they will be better equipped to analyze field problems when they arise. If problems do arise, users should begin by evaluating the entire pumping system to see if there may be a non-pump issue causing the problems. If you find a problem with the pumping system's design or the way the pump is being operated, those issues should be addressed first. If no pumping system or operational issues are identified, then move to the pump. If the pump does seem to be the source of the problem, then the user should perform a detailed pump design audit to determine if a modification to the pump's design is required to obtain the required level of reliability for the application.

Acknowledgement

The author would like to thank Glenn Everett, who is currently a controls supervisor for a natural gas pipeline company, for reviewing the processing portion of this work and making valuable suggestions that helped make it more comprehensive.

References

1. Dutton, J. A., An Overview of Refinery Products and Processes, Petroleum Processing, e-Education, Penn State University, Fall 2020.
2. Bloch, H. B., Selecting the Right Pump for Oil & Gas Applications, *Pumps & Systems Magazine*, 02/05/2020.
3. Perez R. X. and Bloch, H. B., *Pump Wisdom*, 2nd Edition, Wiley Publishing, 2021.
4. Perez, R. X. *How to Select the Right Centrifugal Pump*, pp. 38-41 & 54-56, Author House Publishing, 9/23/2015.

5. Simpson, Jeff, A Beginner's Guide to Pumping Slurry, Crane Engineering Blog, January 6, 2015.
6. Perez, R. X. *Operator's Guide to Centrifugal Pump*, p. 94, Xlibris Publishing, 2008.
7. Steemit.com blog.

7

Practical Ways to Improve Mechanical Seal Reliability

By Robert X. Perez

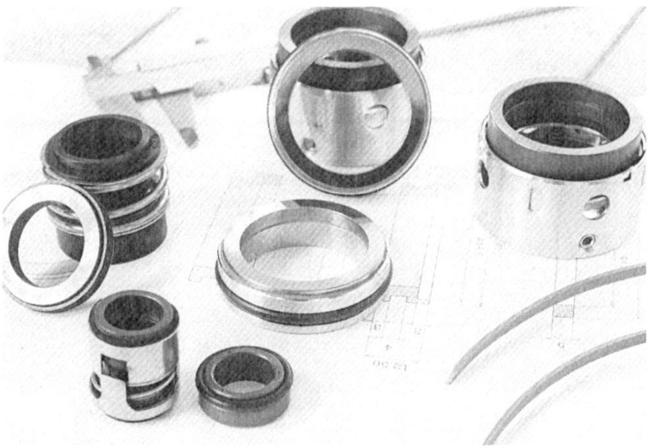

Figure 7.1 Mechanical seal reliability performance has a profound effect on the overall reliability of centrifugal pumps.

Mechanical seals are critical to the success of a centrifugal pump's reliability performance (see Figure 7.1). They must:

1. Prevent leakage to the atmosphere over the expected range of process conditions
2. Provide a reasonable service life
3. Be replaceable and repairable so that there are economical to maintain
4. Be reasonably priced

Robert X. Perez (ed.) Maintenance, Reliability and Troubleshooting in Rotating Machinery, (141–180)

5. Be able to withstand minor upsets and abuse since process variation should always be expected.

Today's mechanical seals have been improved through years of trial and error, design refinements, material improvements, etc. They are marvels of modern engineering. When properly designed, selected, and installed they have been proven to provide exemplary reliability performance. However, for seals applications to be successful, owners must insure they provide manufacturers with a detailed set of design standards and best practices that are read and carefully followed. Any breakdown in the process of designing, selecting, or installing mechanical seal auxiliaries could result in years of frustration due to poor seal performance. This collection of mechanical seal reliability improvement ideas is aimed at helping mechanical seal end users eliminate problem seals through the use of best practice and the analysis of failure data and failed seal components.

Seal Reliability Tracking

To improve mechanical seal reliability, we need to know the current level of reliability and have a reliability benchmark for reference. The mean time between seal failures (MTBSF) is a simple way of determining your average seal reliability. The MTBSF can be trended over time and compared to a benchmark to see if things are improving or getting worse over time.

Before focusing on seal reliability, let's briefly discuss overall pump reliability. The mean time between repairs (MTBR) metric is one of the most widely used metrics for tracking centrifugal pump reliability by pump users throughout the industry. To calculate it you simply need to know the number of failures in a given period of time and the total number of pumps in your population. The equation for MTBR is:

$$MTBR = \frac{M \times T}{R}$$

Where M is the total pump count, T is the reporting time, and R is the total number of repairs during the reporting period. For example, let's say we have 200 pumps in our population, and you have 20 failures in a 3-month period. This means your MTBR is (200 x 3)/20 or 30 months between repairs.

To be useful, MTBR data must be trended over a significant period of time (12 months or more) and compared with similar populations of pumps.

For example, it doesn't make much sense to compare the MTBR for water pumps with those of pumps in hot oil service. Similarly, it doesn't make sense to compare centrifugal pump MTBR with reciprocating pump data.

MTBR Data from Across the Industry

One of the first questions asked by management is: What benchmark should we use for pump MTBR against? There are many factors to consider before answering this question, such as type of industry, pump type (API or ANSI), and average experience level of the operators and mechanics. Well-known author and machinery guru, Heinz Bloch, listed the following MTBF statistics for pumps (Table 7.1) in his March 2006 column in *Hydrocarbon Processing Magazine* as a guideline for pump users:

You can see a great deal of variability in the MTBR results across the industry.

The same mean time between failure concept can be used to evaluate mechanical seal reliability performance. However, for seals we want to know the mean time between seal failures (MTBSF), which is calculated by knowing the number of seal failures in a given period of time and the total number of seals in the population. The equation for MTBSF is:

$$MTBSF = \frac{\sum Seals \times T}{R_{seals}}$$

Table 7.1 Centrifugal pump MTBF statistics.

ANSI Pumps, US average	2.5 years
ANSI/ISO pump average, Scandinavian pulp and paper plants	3.5 years
API pumps, US	5.5 years
API pump, Western Europe	6.1 years
API pumps, repair focused refinery, developing country	1.6 years
API pumps, Caribbean	3.9 years
API pumps, best-of-class, petrochemical plant, US, California	9.2 years
API pumps, best-of-class, petrochemical plant, US, Texas	10.1 years
All pumps, major petrochemical company, US, Texas	7.5 years

Table 7.2 MTBF targets for mechanical seals in refinery services.

Excellent	>90 months
Very good	70 to 90 months
Average	70 months
Fair	62 to 70 months
Poor	< 62 months

* Mechanical seals in chemical facilities, installed in ANSI pumps, can be expected to have average lives that are approximately 50 to 60% of refinerv values.

Where Seals is the total seal count, T is the reporting time, and R_{seals} is the total number of seals replacements during the reporting period. For example, let's say we have 500 seals in our population, and you have 30 failures in 6-month period. This means your MTBR is (500 x 6)/30 or 100 months between seal failures. In the same *Hydrocarbon Processing Magazine* column, Bloch went on to suggest the following MTBR targets for mechanical seals in refinery services*, based on his experience and extensive failure data collected by the John Crane Company (see Table 7.2):

These suggested benchmarks may seem high, but keep in mind these are targets to shoot for. Only by knowing what is possible can we attain world-class performance. Note that some organizations multiply there calculated MTBSFs by a factor from 0.5 to 0.75 to account for the fact that not all pumps are running at the same time. Others use the actual number of installed seals, since about 15% of pumps in petrochemical application are double ended. Considering these complicating factors, it is difficult to come up with an exact value for a site's MTBSF. However, as long as your calculations are consistent, you will be able to determine if your improvement efforts are paying off.

Reliability Tracking Tools

The failure growth plot is a simple but useful tool for tracking the reliability performance of a single or set of similar pump seals. It is constructed by plotting cumulative time on the y axis and the cumulative number of seal failures (or repairs) on the x axis. (The term "cumulative" means you

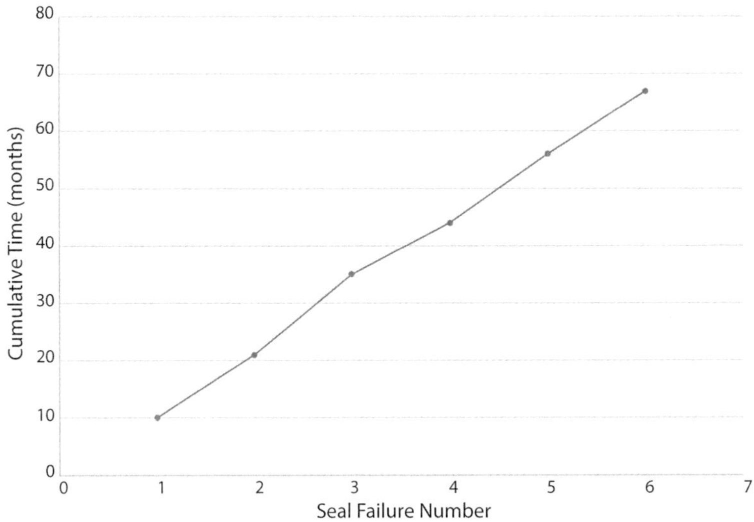

Figure 7.2 A failure growth plot showing a constant failure trend.

are continuously adding up the variable being evaluated. For example, if four seal repairs occur after 10, 11, 14, 9, 12, 11 months of operations, the cumulative values corresponding to these repairs would be 10, 21, 35, 44, 56, and 67 months. The associated cumulative repair values would be 1, 2, 3, 4, 5, and 6 failures.) Figure 7.2 shows a failure growth plot corresponding to this failure progression. Notice that since the failure rate is fairly constant, the failure growth plot appears linear.

Figure 7.3 shows a failure growth plot for three different seals. Notice that one plot shows a fairly constant failure rate, one plot that is improving, and one plot that is deteriorating. By comparison, you can quickly determine if seal reliability is improving or not.

Bad Actors

The best means of improving your site's machinery reliability performance is by identifying and addressing what are called bad actors. Bad actors are simply your worst 5 to 10 pumps, based on the number of times they fail in a given time period. It has been determined that MTBR data following the 80/20 rule, which means that 20% of your worst pumps cause 80% of all the failures in your population. If you think about this for a minute this statement should impress you. The alarming consequence of this statement

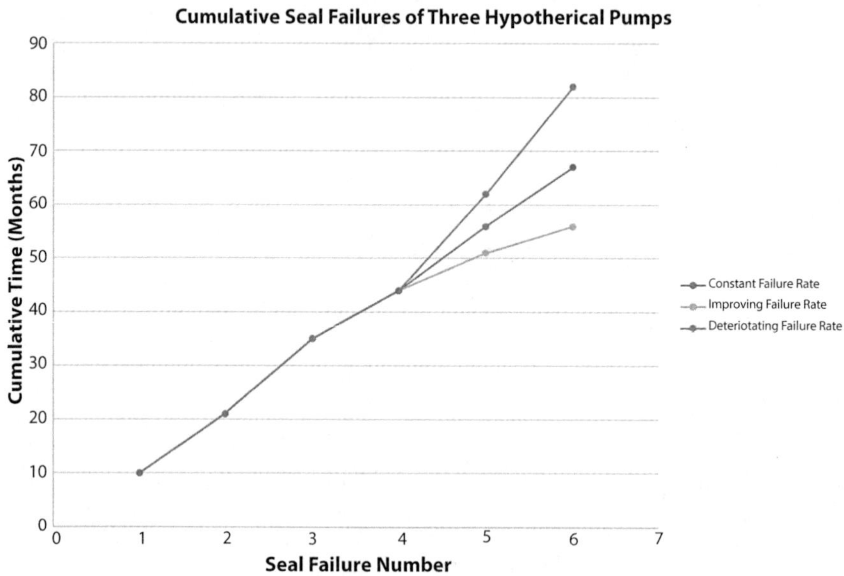

Figure 7.3 Multiple failure growth plots.

is that our maintenance organization spends 80% of its time and money on only 20% of its pumps!

The bad actor concept also applies to mechanical seal reliability. We find that about 20% of the seal applications cause 80% of all seal failures. Since we commonly see that many seal failures are caused by pump design, installation, and misapplication issues, it makes sense to develop a pump-based methodology for analyzing seal failures. Reasons for these failures include [1]:

- Operating problems = 40%
- Mechanical issues, i.e., assembly errors, incorrect clearances, poor alignment, etc. = 25%
- Faulty system design, i.e., insufficient suction, inadequate flush system, poor hydraulic fit = 19%
- Seal component selection inadequate, i.e., wrong materials or configuration = 9%
- Miscellaneous causes = 8%

Here is my recommended list of action steps to take when dealing with troublesome seals:

1. Force rank troublesome pumps for the area of interest
2. Select the top 5 to 10 bad actors to attack aggressively
3. Assign a reliability team to each bad actor. Teams should consist of the following members:
    a. Process engineer
    b. Reliability professional (leader)
    c. Process operator
    d. Mechanic
    e. Safety professional (as required)
    f. Manager (as required)
    g. OEMs (as required)

8. Compile a detailed database of reliability data for each bad actor
    a. Equipment details
    b. MTBR
    c. Outage information, such as length of outage, economic impact of an outage, etc.
    d. Failure plots (reliability growth plots)

5. Whenever a bad actor fails, the responsible team members should be informed immediately. It is vital that the repair shop keep the bad actor team informed of all failures and any new developments, such as pertinent disassembly observations, assembly issues, etc.
6. The reliability engineer should oversee updating bad actor database files as developments occur.
7. The reliability teams should meet regularly, i.e., monthly or quarterly, to discuss failure details and share improvement ideas.
8. Collectively, all the reliability team members should regularly meet with management to present their progress and recommendations to management.

The roles and responsibilities of bad actor team members are shown in Table 7.3. Notice that some members are more involved than others. Some of the members only serve as consultants during the process, while others are simply kept in the loop. This distribution of responsibilities allows the analysis process to be time efficient. When a machine falls off the bad actor list, the team is dissolved, perhaps to join another bad actor team.

Table 7.3 Roles and responsibilities matrix for bad actor team.

Task	Roles						
	Reliability professional	**Process engineer**	**Process operator**	**Mechanic**	**Safety professional**	**OEM**	**Management representative**
Failure notification	I	I	I	I	I		
Root cause failure analyses	L	P	P	P	C	P	I
Recurring progress meetings	L	P	P	P		I	
Outside lab analyses: metallurgical, oil, field studies, etc.	L			P		C	
Internal research: • Internal Lab samples • Gathering process data		L	P				

(Continued)

Table 7.3 Roles and responsibilities matrix for bad actor team. (*Continued*)

Task	Roles						
	Reliability professional	**Process engineer**	**Process operator**	**Mechanic**	**Safety professional**	**OEM**	**Management representative**
Tracking progress • MTBF calculation • Reliability growth plot	L	I	I	I		I	I
Recommendations	L	P	P	P	C	C	I
Follow up	L	P	P	P	C	C	I

L - Lead
P - Participant
C - Consulted as needed
I - Informed

Management's initial role should be to form teams for each bad actor. By forming these reliability strike teams, every bad actor is made a priority and given the due diligence it deserves. Management needs to also track the program's progress and provide resources required to keep the bad actor program moving forward. The hope is that these teams will raise the sitewide awareness of bad actors and provide the resources required to ultimately uncover the root causes of failure.

For this approach to be successful, every bad actor failure must be scrutinized by the team members so that the root cause is uncovered. Only by knowing a failure's root cause can effective design upgrades be proposed. Detailed failure data is required to chart the best path forward. The strike teams should all discuss their findings and then agree on the improvement plan before it is presented to management.

Mechanical Seal Best Practices

To be reliable, a mechanical seal needs to be properly installed and operated within a cool and non-clogging environment. Installation best practices are highly recommended to ensure your mechanical seals are given the best chance to perform as the designer intended. Here is a quick review of some commonly accepted best practices aimed at optimizing the MTBSFs. The following specific best practices are covered:

1. Stuffing box inspections and tolerance limits
2. Seal flush piping recommendations
3. Piping Best Practices for Seal Support Reservoir Systems
4. Seal Cooler Layout Best Practices

1. Stuffing bo.x inspections and tolerance limits

For a mechanical seal to provide an exemplary operating life, we must ensure there is precise seal to seal housing alignment. A precise seal to seal housing alignment means that the shaft sits inside the seal housing concentrically and is perpendicular to the stuffing boxes faces. Furthermore, we need to ensure that a) the shaft is straight, b) there is a proper amount of shaft play or movement, and c) that the bearing housing fits are within limits. Here are all the recommended stuffing box checks and tolerances required to ensure reliable seal performance (see Figure 7.4):

FIG. 7.4.A: Radial shaft movement (shaft runout) 0.0015–0.003 in.
FIG. 7.4.B: Radial bearing fit 0.002–0.003 in.
FIG. 7.4.C: Bearing frame perpendicularity 0.0005 in./in.
FIG. 7.4.D: Axial shaft movement (end play) 0.002 in.
FIG. 7.4.E: Seal chamber bore concentricity 0.005 in.
FIG. 7.4.F: Seal chamber face squareness 0.0005 in./in.

Sealing surfaces of the shaft, sleeve, bore, and seal chamber face must have a surface finish limit of 63 Ra-μin (Ra=roughness average). For ease of installation, the leading edge of the shaft or sleeve should be chamfered as shown and all parts should be deburred.

2. *Seal Flush Piping Installation and Operations Best Practices:*
   a) Tubing should have a minimum diameter of ½ inch (13 mm) for shafts 2-1/2 inches (64 mm) and smaller. Tubing should have a minimum diameter of ¾ inches (19 mm) diameter for shafts 2-1/2 inches (64 mm) and larger. To minimize friction:
      a. Use long, smooth radius bends
      b. Use larger tubing whenever possible
      c. Use rigid tubing whenever possible.
   b) Use thread sealant instead of threat tape, because thread tape can break apart and damage the seal.
   c) Minimize the use of 90° elbows; 45° elbows are preferred.

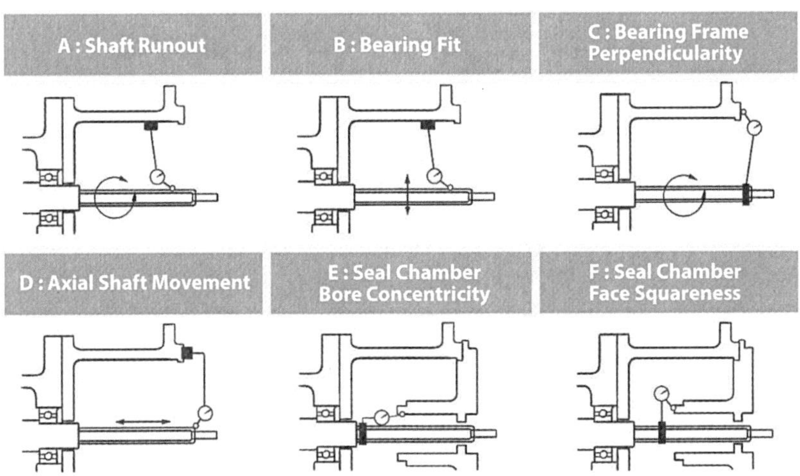

Figure 7.4 Maximum inspection variations (TIR) allowed in the stuffing box.

d) Use continuously rising lines at a minimum slope of 0.5 in./ft (42 mm/m). There should be no relative high points that could trap air in the tubing.

e) For all single or dual unpressurized seals, the product must be a good face lubricant for reliable seal performance.

f) Note: All recirculation flow from the pump discharge to a lower pressure region will result in some loss in efficiency.

3. *Piping Best Practices for Seal Support Reservoir Systems:*

a) The reservoir should have a minimum 3-gallon (11.4 liter) volume for shafts less than 2-1/2 inches (64 mm). The reservoir should have a minimum 5-gallon volume (18.9 liters) for shafts larger than 2-1/2 inches (see Figure 7.5).

b) The reservoir normal liquid level (centerline of the sight glass) should be 36 inches (1 meter) above the gland centerline.

c) The reservoir should be located as close to the pump as possible (within 36 inches (1 meter)) while still leaving room for operation and maintenance.

d) Do not mount the reservoir directly above the pump.

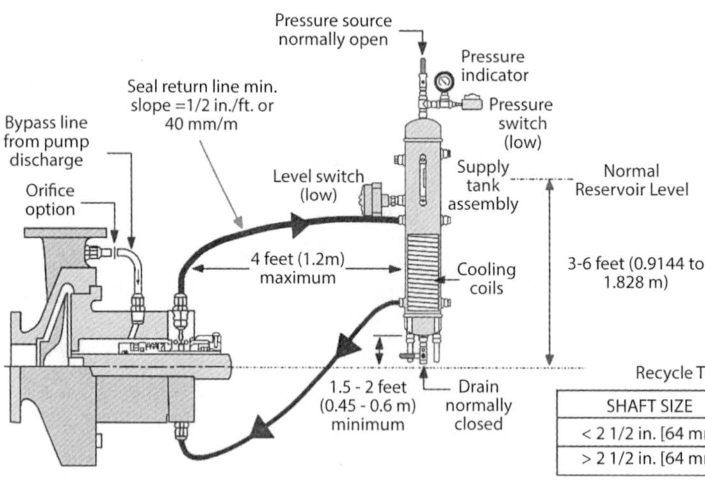

Recycle Tubing Size

SHAFT SIZE	⌀
< 2 1/2 in. [64 mm]	1/2" or greater
> 2 1/2 in. [64 mm]	3/4" or greater

Figure 7.5 Guidelines for seal support reservoir systems.

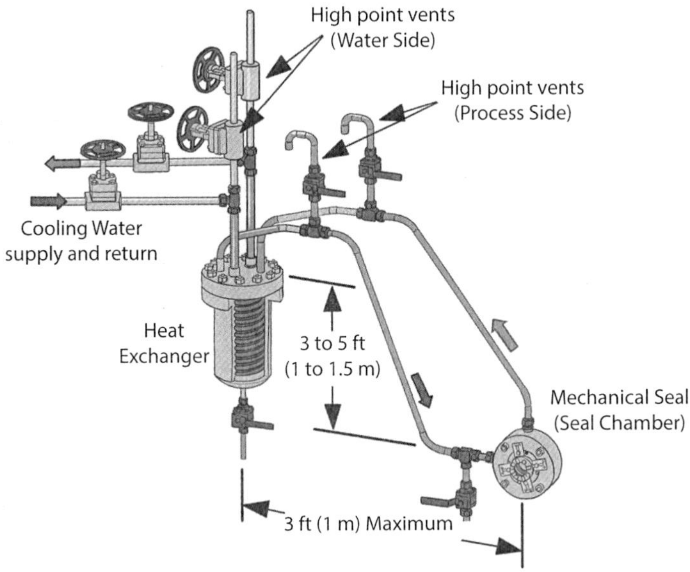

Figure 7.6 Seal cooler guidelines.

4. *Seal Cooler Layout Best Practices*
   a) A major mechanical seal supplier recommends that the seal cooler should be mounted 3 to 5 feet (1 to 1.5 m) above the gland centerline (see Figure 7.6). Check with your seal supplier for their specific cooler installation guidelines.
   b) Coolers should be located as close to the pump as possible (within 36 inches (1 meter) radius) while leaving room for operations and maintenance.
   c) Do not mount the cooler directly above the pump.

Improved Mechanical Seal Support System Designs [2]

The first edition, API 682 Shaft Sealing Systems for Centrifugal and Rotary Pumps (see API Standard 682 insert below). was published in 1994 with the following mission statement: "This standard is designed to default to the equipment types most commonly supplied that have a high probability of meeting the objective of at least three years of uninterrupted service, while complying with emissions regulations." Notably, the standard includes recommendations that, unless otherwise specified, default to technology that has proven to be safe and reliable. Currently in its fourth edition, API 682 continues to offer guidance based on process service for both mechanical seals and their support systems.

API Standard 682 is a standard dealing with the design and application of end face mechanical seals commonly used in centrifugal and rotary pumps. This standard is based on the combined knowledge and experience of seal manufacturers, engineering companies and end users. API 682 is primarily intended for use in the petroleum, natural gas and chemical industries but is often referenced for other types of equipment and industries.

Reducing Potential Leak Points

Throughout API 682 4th Edition, there are references to reducing the number of connections in seal support systems. Whether welded pipe or tubing is selected for the system, threaded systems are discouraged. *Every connection can be viewed as a potential leak point and possible reliability risk in hydrocarbon pumping applications.* Leaks on seal support systems near pumps can cause asset damage, increased downtime, environmental issues, and safety risks.

In the past, many seal support systems were constructed out of pipe due to piping being historically preferred. More recently, seal manufacturers, end users, and pump OEMS have implemented tubing as a connection solution in seal support systems due to its long history of successful use in critical applications throughout the industrial world. As rotating equipment expert Heinz Bloch noted in a March 2018 *Hydrocarbon Processing Magazine* article, "Reliability: Consider stainless steel tubing for mechanical seal connections", "[the] American Petroleum Institute Standard 682 (API 682) began to endorse the use of tubing for some seal piping plans. Regrettably, tradition-bound purchasers still opt for hard pipe; we are asking them to reconsider." API 682 (4th Edition) now specifies both hard pipe and tubing seal support system connections.

Tubing can be utilized to reduce the number of connections by bending lines and appropriately using adapter fittings. Often, the only needed connections are those at the seal and the sealing system. Since tubing is annealed, bending the tubing work hardens the metal, increasing the strength of the tube at the bend. Innovative connection technologies such as flange adapters and extended male connectors further reduce the number of connections from threaded ports on the seal and seal pots by eliminating the need for multiple fittings. The use of tubing provides further financial benefit when we examine the maintenance, repair and operating (MRO) costs of the pump, seal, and support system. During maintenance operations where "piping" around pumps is reworked, the use of tubing eliminates the need for costly on-site welding and can be installed quickly to reduce downtime.

Simplifying Operation and Maintenance

Seal support systems are critical to the proper operation of the seal and pump, and as such, require regular visual inspection. Making the job of visual inspection simple promotes system reliability and safety. When designing seal support systems, there are several best practice design principles to consider once the piping plan and general arrangement have been selected.

1. The systems function should be apparent and the design intuitive.
Mechanical seals are often damaged when pumps are started and stopped, sometimes as the result of improper seal support system operation. If the design of the seal support system facilitates proper operation, common mistakes when commissioning pumps can be avoided.

For example, let's assume an API 682 4th Edition Plan 32 (Figure 7.7) is installed using piping, with the instruments located away from the seal it's

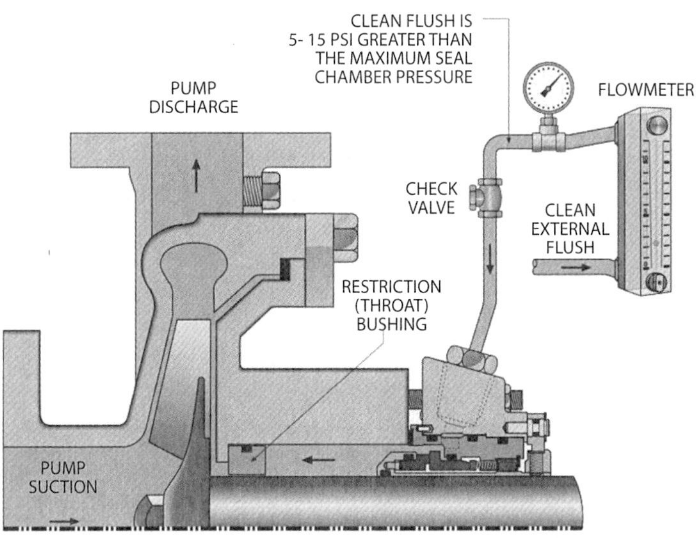

Figure 7.7 API seal flush Plan 32.

protecting. While functionally correct, this design provides the operator little information regarding the operation of the system, what information is important, and why it is important. If the instruments are located down next to the seal on a pump by using tubing, the operator can walk by the pump and bend down to read instrument information. Creating even small obstacles for operators increases the risk that trouble signals will be missed and this reduces reliability. An intuitive solution is to arrange these components on a panel.

Placing a piping plan such as a Plan 32 on a panel accommodates two important design principles: proper identification and operation. This provides the following benefits:

- First, the Plan 32 now looks like a distinct system. Instead of only having components located on a pipe or tube run, they are laid out on a panel where flow path can be identified, and all instruments are available at eye level.
- Additionally, API 682 supports these design considerations. It states: "All controls and instruments shall be located and arranged to permit easy visibility by the operators, as well as accessibility for tests, adjustments, and maintenance."
- Lastly, panels can include part numbering information, flow path indication, and operator instructions. These improvements help ensure safe and reliable startup and shutdown of pumps and seal support systems.

2. The system should be designed for ease of maintenance.

While correct operation of the seal support system is a high priority, consideration should also be given to designing systems that are easy to maintain. Seal support systems contain commonly serviced items such as flow meters, strainers, and other visual instruments. Preventative maintenance (PM) on these systems should be simple and safe for operators. If strainers are not conveniently positioned and available for blowdown, it is unlikely that proper PM will be performed at recommended intervals.

API 682 4th Edition also recommends block-bleed configurations for all gauges. If systems are not designed with this feature, it is likely that as gauges fail, operators will be left without critical information until the next turnaround or project when the pump and support system can be decommissioned, and the gauge replaced.

Lastly, there are a wide variety of tubing connections and design options that allow for every serviceable component on a seal support system to be easily removed and replaced while continuing to operate the system.

For seal pots, the 4th Edition stipulates "Local operation, venting, filling, and draining shall be accomplished from grade. Unless otherwise specified, systems that require the use of a ladder or step or that require climbing on the baseplate or piping are not acceptable." Many plants have older seal pots with just a pipe plug at the top. Having operators climb a ladder to top-off the pot can expose them to process vapors and is a generally unsafe practice. Designing Plan 52s or Plan 53s with fill systems such as the one shown below is a simple design consideration and best practice that promotes safe maintenance.

Building Better Seal Support Systems

Implementing these basic best practice design principles for mechanical seal support system increases reliability and reduces costs. To recap how you can realize better results with your systems:

- Consider using tubing instead of welded pipe to reduce installation and maintenance costs. This will also promote reliability through the reduction of potential leak points.
- Bringing pumps offline to fix minor instrumentation issues or fill seal pots should not be acceptable. Locating these systems on panels, with proper labeling and designing for easy maintenance, reduces the chance for operator error which can damage seals.

Seal failures and the associated costs of seal replacement should be of great concern to rotating equipment groups at all plants. Ensuring that the best practices and design principals of API 682 4th Edition are followed helps prevent these costs and creates a safer and more reliable operation.

Common Mechanical Sealing Design Challenges

Sealing Light Hydrocarbon Liquids

Mechanical seals consist of two extremely flat surfaces that slide against each other as the pump shaft rotates (see Figure 7.8). The surfaces are held in place by either a spring or bellows arrangement and pressed together by the hydraulic pressure of the liquid. A fluid film is generated between the two sealing surfaces, by the pumped product, which prevents them from

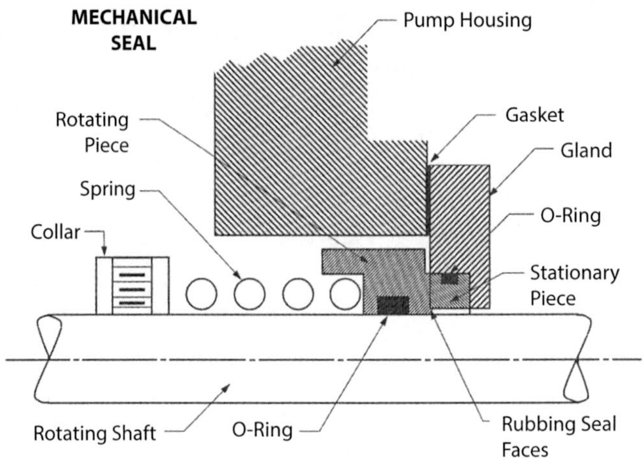

Figure 7.8 Cross section of a pusher-type mechanical seal.

rubbing without lubrication. Heat is generated within the seal due to rubbing of the seal faces and viscous shear of the fluid film between the faces. For low viscosity fluids, most of the heat generation results from the seal faces rubbing together. The absence of this fluid film will result in frictional heat and a rapid destruction of the mechanical seal.

Two major design features that seal designers can tweak to ensure proper face lubrication and seal stability are the seal's balance ratio and the seal's cooling scheme. (The balance ratio of a seal is defined as the ratio of the hydraulic closing area (A_c) to the hydraulic opening area (A_o), customarily expressed as a percentage.) The most common conditions that can cause seals to fail prematurely are a) running a pump dry, b) vaporization of the sealing liquid at the seal faces (see Figure 7.9), and c) excessive shaft vibration.

Some common ways to deal with light hydrocarbon liquids:

- Add cooling to move the liquid's starting point father away from the vaporization line (Figure 7.9) to ensure there will be a liquid film across most of the seal faces.
- Increase pressure inside the seal cavity by using an API Plan 11 and a throat bushing to move the operating point away from the vaporization line (Figure 7.10).
- Isolate primary seal by using a double seal with buffer liquid to completely avoid the possibility of flashing

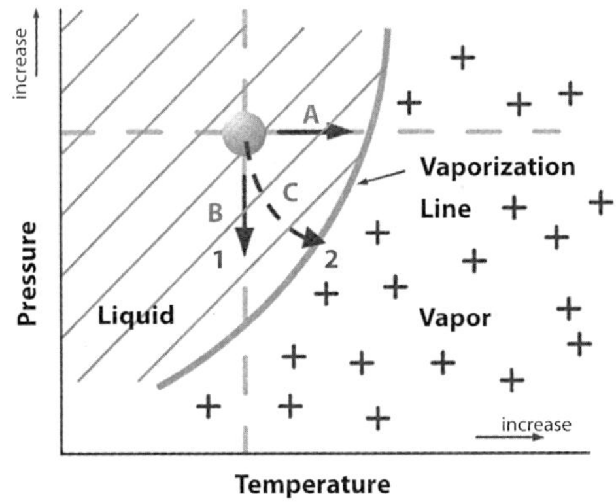

Figure 7.9 The red curve is a vapor pressure curve for a hypothetical liquid. To the left of the red curve, the fluid is in the liquid state and to the right of the red curve the fluid is in the vapor state. Path "B" above is a hypothetical leakage path across sealing faces where no temperature increase occurs as it reaches point 1 (atmospheric pressure) in the liquid state. In contrast, Path "C" shows seal leakage that gets heated up as it travels to point 2 (atmospheric pressure). Notice in the Path "C" case that, because of the rubbing of the seal faces, the leakage reaches the vaporization line and begins to flash. It is common for flashing to occur between the seal faces when handling low vapor pressure fluids. For reliable performance, most seal designers recommend that there needs to be liquid across more that 50% of the seal face under operating conditions. Users should try to stay well above the 50% minimum design value for liquid coverage of the seal faces to ensure a stable liquid film is maintained during process upsets or if cooling degradation occurs over time.

- Use a special seal design to handle flashing, such as a wavy face or hydro-pad face design.
- Install tandem seals with an intermediate cavity that pipes primary league to a flare.

Sealing Hazardous Organic NESHAP Liquids

In 1992, the EPA decided to regulate the emissions of certain organic hazardous air pollutants from synthetic organic chemical manufacturing industry (SOCMI) production processes. The EPA decided on the rule, which is referred to as the hazardous organic NESHAP or the HON. The HON requires potential sources to achieve emission limits reflecting the application of the maximum achievable control technology. The rule regulates the emissions of 112 of the organic chemicals due to equipment leaks.

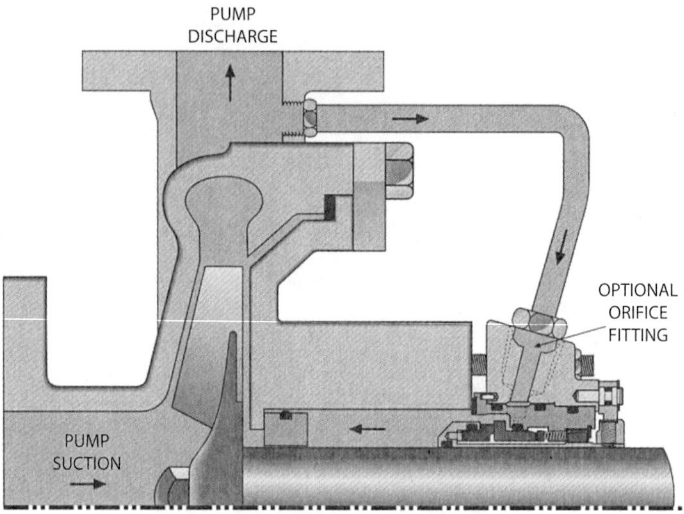

Figure 7.10 API flush Plan 11.

Dual seal designs are often selected to limit the leakage of HON chemicals to the atmosphere.

Buffer Gas Absorption

Note: API 682 cautions against using gas pressurization in direct contact with the barrier fluid when the reservoir pressure is above 150 psig (10 barg). The concern is that gas can be absorbed into the barrier fluid and then flashed off at lower pressures. However, based on field experience and tests, some seal manufacturers have stated that direct contact gas pressurization can be used up to 300 psig (21 barg) as long as the barrier fluid temperature is less than 250° F (120° C).

Excessive Solids

An API Plan 12 (Figure 7.11) seal filtration system can be used in critical applications where large volumes of solids are expected. If properly sized, a large duplex filter (Figure 7.12) can be employed to remove solids from the seal flush stream before entering the stuffing box and plugging up the seal or potentially damaging the seal faces.

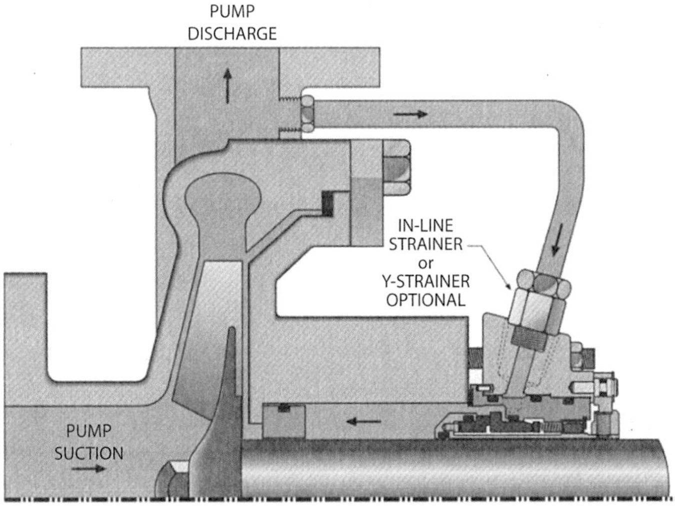

Figure 7.11 API seal flush Plan 12.

API Plan 12 filtration systems utilize a duplex filter cartridge filter design that allows for continuous flow during flow transfers from one filter to the other. These duplex filter cartridges with double block and bleed capability for isolation and online element replacement can be installed so that

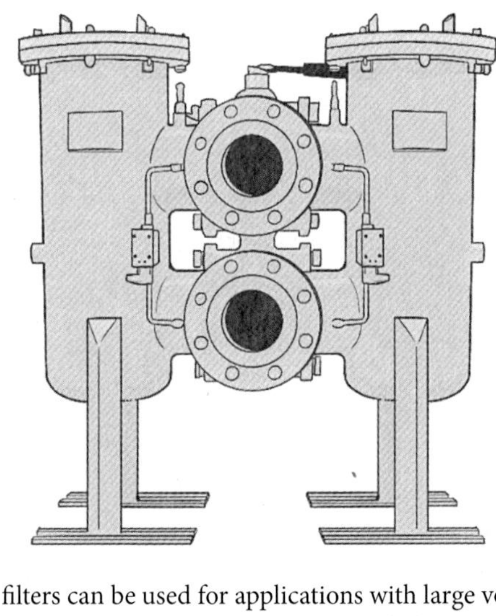

Figure 7.12 Duplex filters can be used for applications with large volumes of solids. Note: An integral magnetic separation system can be used along with a duplex filter arrangement in application where pipe scale may be present due to corrosion.

operators can quickly switch from one filter housing to the another with a single lever, without ever interrupting the seal flush flow.

Seal Cooler Issues in Hot Applications [3]

Pumps moving high-temperature fluids require special considerations to ensure mechanical seal operate reliability. High temperatures can result in fluid properties that can limit the life of a mechanical seals, such as low fluid viscosities, high vapor pressures, and thermal breakdown of hydrocarbons, i.e., coking. Designers know that reducing the fluid temperature in the seal chamber can be beneficial and can result in improved seal performance and equipment reliability.

There are several standardized seal flush piping plans defined in API-682 that are designed to achieve the desired cooling. While both flush plans can reduce the temperature, there are significant differences in their cooling efficiency and impact on the plant operations.

Piping Plan 21

Plan 21 is defined as a circulation of process fluid from a high-pressure region of the pump, through a seal cooler and flow control orifice, back into the seal chamber (see Figure 7.13).

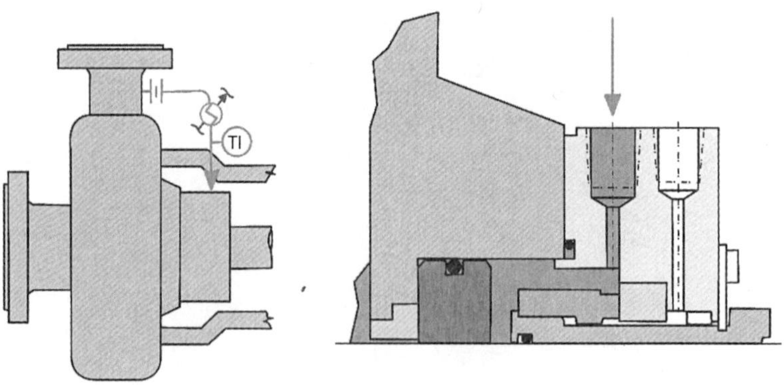

Figure 7.13 Piping plan 21.

Advantages

Since the process fluid is circulated through a cooler, the fluid injected into the seal chamber will be lower than the process fluid. The flow rate of the fluid and desired reduction in temperature will dictate the type and capacity of the seal cooler in the system. Both water-cooled and air-cooled seal coolers have been used successfully in the field.

The Plan 21 configuration creates circulation due to the differential pressure across the piping plan. For this reason, there are generally no issues with overcoming friction losses in the piping and the seal cooler itself. This allows for the selection of a greater variety of seal coolers and more flexibility in the design of the system.

Disadvantages

The greatest challenge in applying a Plan 21 is providing sufficient cooling to satisfy the seal requirements under process conditions. The process fluid enters the cooler at process temperature and must be cooled to the desired temperature across the cooler. This continual removal of heat can require a larger capacity seal cooler and higher demand for cooling water from the plant. Depending upon the actual field conditions, it may also be difficult to achieve the desired temperature reduction. Furthermore, in a Plan 21, the cooled process fluid is recirculated back into the pump casing and the process, which tends to reduce the process fluid temperature in the pumping system. Although this is often considered to have a negligible impact, it does result in loss of the processing unit's thermal efficiency.

The potential for cooler fouling is always a major concern when using an API Plan 21 in high-temperature applications. Scale deposits form when the solubility of dissolved minerals in the cooling water drops due to higher water temperatures, which in turn results in the precipitation of dissolved minerals as scale deposits. Generally, scale deposits exhibit inverse solubility with temperature; therefore, as the water temperature increases, the solubility of scale decreases and scaling becomes more likely. Since the process fluid enters the seal cooler at the maximum temperature of the process in a Plan 21, the highest potential for fouling exists near the cooler's inlet. Poor quality cooling water can accelerate the rate scale buildup. Over time, the scale buildup creates internal blockages, which limits water flow and reduces heat transfer performance.

The maximum application temperature for coolers installed in Plan 21s varies based on the plant's cooling-water quality. Site experience usually dictates what the upper limit for the inlet cooler temperature should be for an acceptable seal cooler life. As a starting point, it is best to limit API 21 piping plans for clean, high-temperature fluids and hot water operating under 80°C (176°F). Adopting this upper limit for cooler inlet temperatures should ensure a reasonable cooler life assuming a typical cooling water quality.

Cooler inlet and outlet temperatures, on high temperature seal applications, should be monitored regularly in the field to determine if fouling is occurring or has occurred. The installation of a temperature indicator in the process piping downstream of the cooler (See "TI" element in Figure 7.13) makes it easier to monitor cooler performance. Furthermore, there should always be a supply of new or refurbished seal coolers in the warehouse in case fouling has affected cooler performance in the field.

Piping Plan 23

Plan 23 is defined as a circulation of process fluid from the seal chamber, through a seal cooler, back into the seal chamber (Figure 7.14). A bushing is required in the back of the seal chamber to minimize the interchange of the lower temperature process fluid in the seal chamber and the higher temperature fluid inside the pump casing.

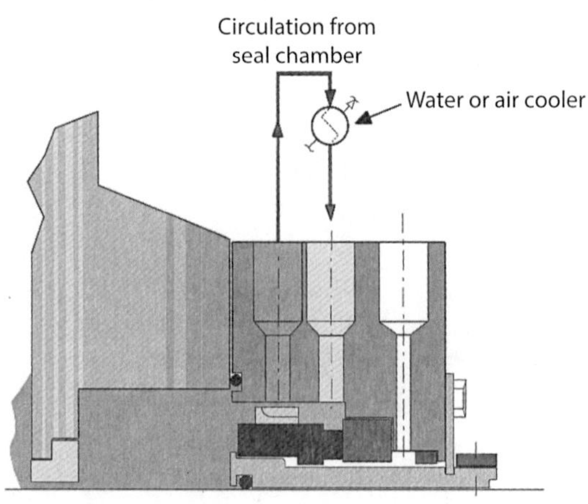

Figure 7.14 Detail of seal piping plan 23.

Advantages

Since the cooled process fluid is recirculated within the piping plan and not lost back into the pump, the Plan 23 can more easily create a lower temperature environment inside the seal chamber. The seal cooler primarily removes heat entering the seal chamber from seal generated heat and heat soak through the pump casing. This means there is a significantly lower cooling requirement for the Plan 23 when compared to the Plan 21. The benefits of the Plan 23 are a lower seal chamber temperature and less heat loss from the pump and process.

The lower heat loads in a Plan 23 can create additional benefits: 1) The lower cooling capacity requirements can allow for smaller seal coolers to be selected and have reduced cooling water requirements. 2) The lower operating temperatures in the piping plan also reduces the potential for fouling in the seal cooler.

Disadvantages

Fluid circulation in a Plan 23 is created by a pumping device on the rotating seal. While these pumping devices (or pumping rings) can create flow, they cannot overcome significant friction losses in the piping system. For this reason, the system is designed with features such as large diameter tubing, a minimum number of fittings, smooth bends, and short piping runs.

A Plan 23 is also a "closed" system. Fluid circulates around from the seal chamber to the seal cooler and back without a natural means for venting gases or vapor from the system. All Plan 23s must be designed with a high point vent (or vents) to fully remove all vapor from the seal chamber, piping, and seal cooler prior to operation (see "High point vents" in Figure 7.6). Process fluids which continually produce vapors may require additional periodic venting while the system is in operation.

Common Considerations for Flush Plans

Thermosyphoning: Similar fluids at different temperatures will have different densities. This density difference can be exploited in seal flush plans to produce a natural circulation flow, called thermosyphoning. In a properly designed Plan 23 circulation loop, the thermosyphon effect can generate more flow than by simply using a pumping ring. To gain this additional seal cooling benefit, the seal cooler must be mounted at an elevation slightly higher than the seal. Even when the pump is on standby, thermosyphoning can produce enough fluid flow to lower the seal chamber temperature.

Heat soak: Since the seal chamber is at a temperature lower than the process temperature in the pump, heat will be transferred into the seal chamber. There are design practices which can reduce heat soak. All seal chambers must have an effective bushing at the back of the seal chamber. This will help prevent intermixing of the high- and low-temperature fluids in this region. The bushing can also be designed to reduce heat transfer through conduction (by using materials with low thermal conductivity) and convection (by reducing the heat transfer area in the seal chamber). Seal glands can also be designed with thermal breaks to reduce conductive heat transfer.

Seal coolers: Seal coolers can be either water-cooled or air-cooled. While both options could be used in either the Plan 21 or 23, in practice one will generally be a better choice depending on the application conditions. In cases where cooling water is not available, the air-cooled seal cooler is the only practical option. In very high temperature applications with a Plan 21, air-cooled may be a better option since it is not prone to fouling. In moderate temperature applications, water-cooled seal coolers can provide lower temperatures. Water cooling also relies on the cooling water temperature in the plant and therefore provides more stable temperatures than an air-cooled unit relying on ambient temperatures.

Fluid properties: It is common for seal OEMs to perform calculations estimating the steady-state temperatures of the Plan 21 and 23. This is critical for activities such as sizing the seal cooler or defining cooling water requirements. It is less common for seal OEMs and end users to consider off-design application conditions such as standby services. For example, if a pump with an air-cooled Plan 23 is on standby in the middle of winter, the design must ask: 1) Will the process fluid solidify in the cooler? or 2) Will the Plan 23 be able to reestablish circulation once the pump is restarted? It is critical to consider all potential applications and upset conditions in the design analysis.

General Seal Piping Plan Recommendations

When it is necessary to reduce the temperature in the seal chamber, either seal flush 21 or 23 piping plans may be applied. In most cases, a Plan 23 will provide the most effective and efficient cooling and would be the first choice. Water-cooled seal coolers also are more effective and will provide more stable operating conditions for the seal. There are, however, applications where a Plan 21 or an air-cooled seal cooler would be more beneficial. End users should consult their seal OEMs to properly engineer a system for their specific set of conditions.

Ways to Improve Seal Reliability Performance

The first step to evaluating a seal's reliability performance is reviewing its MTBSF and plotting its failure growth plot. Is the MTBSF acceptable? Is the seal failure rate stable, increasing or improving? It is crucial to understand how a seal is performing before beginning a detailed analysis. Compare the current performance with the rest of the seal population to get a feel for whether it's acceptable or not.

If you find that the MTBSF is unacceptable, you should have the manufacturer perform a design review, which includes:

1. Reviewing the seal materials of construction
2. Verifying the temperature and vapor pressure of the liquid being sealed
3. Checking the solids content of the liquid being pumped
4. Evaluating the current seal flush plan selection
5. Evaluating the seal flush plan installation in the field to ensure it is installed properly
6. Reviewing the start-up procedure for the pump and seal
7. Checking past repair data on seal pre-installation checks
8. Checking past alignment records
9. Checking current pump operating conditions to ensure they haven't changed since the original installation
10. Checking the current operating conditions, which includes determining vibration and temperature levels and determining the operating condition of the pump (%BEP).
11. Performing detailed RCFAs to better understand the seal failure modes.

Seal Failure Analysis [4]

Performing a detailed seal failure analysis can be helpful in diagnosing, correcting, and preventing repetitive seal failures. Note that there is an important difference between performing a seal failure analysis and troubleshooting. Troubleshooting focuses on immediate problems, while seal failure analyses focus on the specific telltale clues observed during the tear down and inspection of the failed seal.

The goal of the seal failure analysis is not to correct the specific failure, but to correct the condition or design flaw that caused the failure. The first step in performing a failure analysis is finding a large clean table in a well-lit room. Next, lay all the parts on a table with the seal drawing. Make sure to remove all the O-rings so that they can be carefully examined. Then examine all the parts carefully.

There are four fundamental questions that should be asked during the tear down and inspection:

1. What does the seal damage look like? Is it chemical, mechanical, or thermal?
2. How does the damage effect the seal performance?
3. What does the seal damage indicate about the seal's design and operation conditions?
4. What corrective steps should be taken to prevent the observed damage from recurring?

Common Seal Failure Modes

Here are the failure modes commonly found during mechanical seal shop inspections:

Overall Chemical Attack, which leaves parts appearing dull, pitted, eroded, or falling apart. Corrosion of seal components results from using the wrong materials of construction or failing to use a barrier fluid to isolate the seal from the corrosive fluid.

Remedial action can include: 1) obtaining a complete chemical analysis of the liquid being sealed and then upgrading the materials of construction to meet the chemical requirements, or 2) using double seals or flushing a single seal with an external source with a clean fluid that's compatible with the process to isolate the seal from a potentially corrosive environment.

Fretting Corrosion, which causes leakage at the secondary seals and damages the shaft sleeve directly under the secondary seal. The site damaged due to fretting, which appears pitted and shiny bright, results from the repeated back-and-forth motion, due to assembly issues and/or vibration, of the secondary seal over the shaft sleeve.

Causes of the excessive vibration of secondary seals are eliminated by making sure shaft runout, deflection, and end play are within recommended limits. Protective hard-faced alloys or coatings can be applied at the location of secondary seals. If possible, TFE, V-rings, wedge rings can be replaced with more forgiving O-rings. Another design option is replacing pusher-type seals with non-pusher type seals that have static secondary seals.

Damaged elastomers, which appear swollen, hardened, bubbled, broken, or blistered, result from the chemical attack or excessive temperatures. The most likely causes are incorrect material selection or barrier fluid contamination or loss.

The fluid being pumped should be chemically analyzed and the elastomer material selection reevaluated. Frequently, the presence of trace contaminants, originally overlooked when specifying the seal, is the cause of the problem. If a suitable elastomeric material cannot be found, then the seal must be isolated using an external seal fluid or replaced with a seal design using a static secondary seal.

O-rings that appear cut or peeled may have been extruded. Causes of this type of damage may be due to excessive pressure or temperature, chemical softening, or improper shaft and groove sizing. The cure can be 1) a reduction in pressure or temperature, 2) change in the seal design, O-ring material or durometer, or 3) proper sizing of the shaft or groove on the machinery.

Leaching, which gives tungsten carbide and ceramic faces a dull and matted appearance, can reduce surface hardness by 5 points or more on the Rockwell A scale. To correct the problem, upgrade the seal base material. Another way to eliminate the problem is to provide a seal arrangement that provides a buffer fluid at the seal faces, such as a single seal with an external flush or a double seal with a suitable barrier fluid.

Face distortion, which results in a non-uniform face wear pattern, frequently causes excessive leakage. Improper assembly or uneven gland nut torquing can result in nonuniform face loading. Other causes include: 1) improper cooling induce thermal stresses and distortions, 2) poor surface finish at the face of the seal housing, due to corrosion or mechanical damage, and 3) driver to pump misalignment.

Seal faces could be re-lapped to remove face distortions. Other corrective actions include using flexibly mounted stationary seal faces to compensate for gland distortion, checking the driver to pump alignment, and retorqueing the gland to recommended values.

Seal face deflection, which appears as an uneven or narrow wear pattern, can be either concaved or convex. Seals with either condition are not stable under cyclic pressure conditions. Causes may be improper stationary seal face support, swelling of secondary seals, excessive pressure, inadequate balancing of hydraulic and mechanical loads on the primary seal faces.

Corrective and preventive measures include checking the seal design's operating limits and consulting the seal manufacturer about other seal designs, considering a flexible mount for the stationary seal face and replacing carbon seal faces with seal faces having a high modulus of elasticity, such as silicon carbide.

Tracking, which is a wear pattern on the larger face that is wider than the thinner face, is caused by misalignment, pipe strain, cavitation, poor pump condition, or excessive pressure. The problem is corrected by checking and adjusting the seal in its housing and ensuring operating and machine conditions are correct. In instances where seal face alignment is not possible, a design with a static secondary seal may provide enough compensation.

Heat checking is the presence of fine-to-large cracks that radiate outward from the center of the seal ring. These cracks act as a series of cutting and scraping edges against the opposite seal face. Heat causes this condition due to a lack of cooling, vaporization at the seal faces, or excessive pressures or face velocities.

Temperatures can be reduced by flushing or cooling the seal faces and eliminating overloading.

Blistering, a symptom characterized by small circular sections that appear raised on carbon seal faces. Blisters separate the seal faces during operation and cause high leakage rates. Blistering commonly occurs in seals in high viscosity applications that are started and stopped frequently.

Possible actions in high viscosity applications are either increasing the sealing temperature or substituting a less viscous buffer fluid. Other actions include eliminating frequent starts and stops or substituting a non-porous seal face material, such as tungsten carbide.

Other pusher seal issues: There are typically a number of insightful observations that can be made during the inspection of a failed pusher type mechanical seal. Each observation provides a clue to an underlying cause of failure. Here is some advice related to the common damage seen during pushing seal inspections [5].

Symptom: **Pins, drive lugs and drive slots are worn.** The drive components have flats or indentations worn into them from the other sealing components.

Causes:

- Poor lubrication creating excessive friction between the seal faces.
- Axial or radial movement causes wear on the drive components during operation.
- Poor equipment conditions can cause excessive movement of the drive mechanism.
- Wear can be caused by poor perpendicularity or concentricity of the shaft to the seal chamber.
- Excessive torque on sealing faces can also cause heavy loads on drive components.

Advice:

- Review equipment process conditions to eliminate off design operation.
- Perform mechanical equipment integrity and vibration checks to ensure they are within manufacturer specifications

Symptom: **Pins and drive lugs are bent or broken.** The pins and drive lugs used to provide torque to the rotating seal face is deformed, bent, sheared or damaged.

Causes:

- Excessive torque due to over-pressurization.
- Poor lubrication creating excessive friction between the seal faces.
- The process fluid freezing or solidifying in the seal cavity, causing adhesion between the seal faces.
- Poor equipment condition causing excessive movement of the seal and overloading the drive mechanism.

Advice:

- Review application and seal selection to ensure you have the right design for the operating conditions.
- Maintain adequate vapor pressure margin.
- Review startup and shutdown procedures to ensure against adverse seal conditions, such as cold startups, running dry, and cavitation.
- Perform mechanical equipment integrity and vibration checks to ensure they are within manufacturer specifications

Symptom: **Rubbing**, which appears as unintended contact between stationary and rotating components.

Causes:

- Shaft misalignment or concentricity issues.
- Bent shaft or large shaft deflection.
- Non-piloted gland ring installed eccentric to the shaft.
- Throat bushing improperly retained to prevent axial movement.

Advice:

- Perform mechanical equipment integrity checks for radial deflection, shaft run-out and concentricity.
- Perform proper seal installation for a non-piloted gland to ensure concentricity with the shaft.
- If possible, convert to a piloted gland ring or shaft centered design.
- Mechanically retain the throat bushing.

Symptom: **Buildup of solids**, which can be caused by:

- Solid deposits hang up the flexible seal components.
- Debris on the seal components that prevent axial motion of the seal faces.
- Sealing integrity is compromised due to solids accumulation at the sealing surfaces.
- Accumulation of solids in the seal chamber prevents the circulation of process fluids around the seal.

- This can occur on the product side or the atmospheric side of the seal.

Causes:

- In high-temperature hydrocarbon applications, coking (which is a heavy sludge-like residue that results from a reaction with oxygen or chemical breakdown) can hang up dynamic seal parts.
- In applications with dissolved solids, salting can occur on the atmospheric side of the seal due to evaporation of the process fluid.
- In thermal-sensitive fluids, weepage can solidify at lower temperatures outside of the process stream.
- In cryogenics or low-temperature applications, environmental moisture can freeze, hanging up dynamic seal parts.
- Seal components exposed to process fluids with suspended solids become packed in services such as minerals, slurries or paper stock.

Advice:

- Change to metal bellows seal design, eliminating dynamic gaskets.
- Control temperature using jacketed seal chambers and/or seal gland. Use heating or cooling as required.
- For freezing, salting or coking, use a clean quench (steam for coking, methanol quench for freezing, API Plan 62).
- Use dual seals with pressurized barrier fluids (API Plan 53 or 54) to control fluid film and minimize seal exposure to solids.
- Add a clean external flush (API Plan 32) to single seals to minimize seal exposure to solids.

Symptom: 1) Breakage or permanent deformation of the springs. 2) Corrosion of springs, causing weakening.

Causes:

- Excessive stress due to seal misalignment or over-compression.

- Spring rubbing in the spring pocket due to excessive cycling or vibration.
- Metal degradation due to exposure to incompatible chemicals.

Advice:

- Correct equipment misalignment.
- Check seal setting and installation procedures.
- Check process chemical compatibility with spring material of construction.

Symptom: **Coking**, i.e., oxidation, which is a black, hard, grainy material buildup on the atmospheric side of the seal face. Coking is a buildup of thermally decomposed hydrocarbon that is converted to a varnish, lacquer or abrasive sludge.

Risks associated with coking:

- Heavy coke buildup can decrease the unit loading on the seal faces and cause excessive leakage.
- Coking can cause hang-up of the dynamic seal face.

Advice:

- Reduce seal cavity temperature by running Plan 21, 23 or 32.
- Remove the oxygen from the atmospheric side of the seal faces by adding a quench Plan 62:
    - A dry steam quench can be used to displace oxygen and wash away any leakage residue that could cause the outer seal face to hang-up. WARNING: wet steam may cause seal problems.
    - A nitrogen quench can be used to displace oxygen but will not wash away residue.

Seal Failure Inspection Notes

As you inspect the seal parts in the shop, take detailed notes that describe the damage you see. Observations should include the part and the damage seen. For example, you may write the primary seal carbon face shows signs of blistering, or the stationary face O-ring is swollen. At this time, the more detail the better. Make sure to keep all the damaged parts until you have finished your report.

Possible Causes

You could list some possible causes and preliminary recommendations, knowing they may change later. For example, you may write: Excessive solids: we may need better filtration.

Meeting with Manufacturer

If there is reason to question the seal design, sit down with the seal manufacturer and go over your findings and beliefs. Listen to the manufacturer's thoughts and beliefs with an open mind. Together, you can decide the best course of action.

Writing the Seal Failure Report with Recommendations

Finally, write your failure analysis report with recommendations. All the recommendations must be based on actual inspection findings. Start out with simple changes such as additional cooling or flush and work your way up to more costly design changes.

Post-Analysis Activities

Always follow up to see if your seal modifications were effective. The only way to determine if you made an improvement is to see if the time between seal failures has increased. Update the failure growth plot for the seal to see if things are improving. If you find the seal is still failing at the same rate, then continue to perform detailed failure analyses until you find the right design or your application.

Justifying Seal Upgrades

In the real world, there is a wide distribution of time intervals between seal failures due to the many factors that affect seal reliability, such as the properties of the fluid beings sealed, the pump's design and condition, and how the pump is operated. Although not all seal failure intervals fall precisely in a normal distribution, this is probably a good first assumption for a large population of seals in different applications.

In a hypothetical normal failure distribution of the time intervals between seal failures, we normally see there is generally a wide variation in seal run lifetimes. There will be a mean or average failure life, low hanging fruit that fall 2 sigma, i.e., 2 σ, below the mean and exemplary performers that fall 2 sigma, i.e., 2 σ, above the mean. The point here is that there is always a category of seals that fall in the low hanging fruit category that represent about 2%, which often have obvious root causes. These are seals we should attack first.

The payback method is a simple one of defining the economic benefits of a seal upgrade to management. The payback period is defined as the initial investment divided by the annualized payback. For a seal upgrade, we determine the payback period as the total costs associated with seal upgrade divided by the annualized benefits, i.e., Payback period = Installation cost/ Annualized benefits.

Installation costs can include, but is not limited to, the following items:

- Cost of new seal or seals
- Installation costs
- Warehouse spares
- Auxiliary seal plan components required for the upgrade seal

The annualized benefits may include:

- Reduction in repair costs
- Reduction in production losses

Note that annualized benefits are relative terms. To define an economic benefit, you need to have a future annualized repair costs and base annualized repair costs. For example, if a seal is failing twice a year and costs $10,000 per repair, then the base annualized repair cost is $20,000 per year. If you expect the seal failure rate interval to increase to once every 5 years as a result of an upgrade, then the future annualized repair costs should be $10,000/5 or $2,000 per year. In this example, the annualized benefits of the upgrade are expected to be $18,000 ($20,000-$2,000) per year. The point here is you must always consider the future case costs and the base case costs in determining the annualized benefits, i.e.,

Payback period = Total Cost/DRepair Costs = Total Cost/(Base Repair Costs - Future Repair Costs).

Note: Refer to Table 7.4 for guidelines on project paybacks and the likelihood that the project will be approved by management. This guidance is based on the author's personal experiences trying to justify upgrades.

Let's assume the base seal MTBF is $MTBF_A$ and the target MTBF is $MTBF_T$. Then the annualized benefit of an upgrade will be:

$$Annualized\ Benefits = \left(\frac{1}{MTBF_A} - \frac{1}{MTFB_T}\right) \times R$$

where R is the cost of a seal repair.

If we define $1/MTBF_A$ as 1/A and $1/MTBF_T$ as 1/T, then the annualized benefit becomes:

$$Annualized\ Benefits = \left(\frac{1}{A} - \frac{1}{T}\right) \times R$$

Plugging this relationship into the formula for determining the payback period we get:

$$Payback\ Period = \frac{C}{\left(\frac{1}{A} - \frac{1}{T}\right) \times R}$$

To simplify, we must find the common denominator of (1/A-1/T) and combine terms:

Table 7.4 Likelihood project will be approved, based on the payback period.

Payback period range	**Comment**
< 1 year	Sure to be approved
>1 year but <2 years	Likely to be approved
>2 years but <5 years	Difficult project to sell
>5 years	Not likely to be approved

$$Payback\ Period = \frac{C}{\left(\frac{T}{AT} - \frac{A}{AT}\right) \times R}$$

$$Payback\ Period = \frac{C}{\left(\frac{T-A}{AT}\right) \times R}$$

Next, we invert the (T-A)/AT to arrive at the following form of the formula:

$$Payback\ Period = \frac{C}{R} \times \left(\frac{AT}{T-A}\right)$$

Simplifying further, we obtain a simplified payback formula to evaluate seal upgrades:

$$Payback\ Period = \frac{CAT}{(T-A) \times R}$$

where:

C = Cost of the seal conversion or upgrade
A = Actual MTBSF in years or months
T = Target MTBSF (in years or months). As a first guess, you can use the mean seal failure rate of the population as the target design value.
R = seal repair cost. This value should include all the costs associated with a seal repair. For critical pump applications, the term R* can be used to denote that repair costs and production losses due to seal repair outages are being included in the payback calculation.

Readers should keep in mind that this handy formula should be considered a first pass test for determining the viability of a possible seal upgrade and should not be considered the last word on justification. If the payback results are promising, more detailed economic evaluation should be conducted.

Example #1

Assume: C=25000, A=0.5 years, T=5 years

$$Payout\ \ period = \frac{CAT}{(T-A)\times R} = \frac{\$25000 \times 0.5\ \ years \times 5\ \ years}{(5-0.5)\,years \times \$10000} = 1.389\ \ years$$

Comparing this payout period result to Table 7.4 guidelines, we see that this seal upgrade is likely to be approved.

Example #2

Assume we have a centrifugal pump that has a mean time between failure of 7 months and costs $8,000 to replace when it fails. Let's also assume the average seal MTBF for the site is 4.5 years. Next, we need to determine the cost of the seal upgrade. You have determined the cost of new seals, flush plan modifications, and spare seals will be $21,000. Plugging in these values on our equation we get:

$$Payout\ \ period = \frac{CAT}{(T-A)\times R} = \frac{\$21{,}000 \times 0.583\ \ years \times 4.5\ \ years}{(4.5-0.5)\,years \times \$8{,}000} = 1.722\ \ years$$

Comparing this payout period result to Table 7.4 guidelines, we see that this seal upgrade is likely to be approved.

Closing Thoughts

Mechanical seal technology is a complex field of study. Some seal specialists have spent their whole lives acquiring and applying mechanical seal expertise across the industry in hopes of providing the best design for each application. However, in some cases, even veteran designers may have had to tweak their designs to solve difficult seal applications before hitting on one that works. Trial and error is often part of the seal reliability improvement process.

There are no shortcuts to improving the reliability performance of a problem seal. The process involves gathering data, observing damage, determining the root cause of the observed damage, and then testing potential design improvements to see what works. Remember to look at the whole system, not just the seal. Small changes in the seal's operating environment can have a significant effect on the seal's performance.

I have found that it often takes a team to solve the most challenging application issues. Every team member brings a slightly different insight to the problem. Process engineers understand the process, mechanical engineers understand pump and seal design details, the operator knows the pump's operating history, and the mechanic brings repair history to the analysis. Together, the team can identify the root cause of the failure and make the right improvement required to obtain exemplary performance.

References

1. Wallace, N. M, and David, T. John, Pump Reliability and Improvement through Effective Seal and Coupling Management, Texas A&M University, *Proceedings from the 15th International Pump User Symposium*, pp 27-39.
2. Hunsicker, S. Building Better Mechanical Seal Support Systems, Swegelok.com website.
3. Huebner, Michael, *Pros & Cons of Mechanical Seal Piping Plans: Plan 21 and Plan 23*, Empowering Pumps Website.
4. Foszcz, Joseph, L., Senior Editor, *Preventing Mechanical Seal Failures*, Plant Engineering Magazine, Sept. 1992, pp. 47-51.
5. Flexachem, Your Guide to Self-Diagnose Mechanical Seal Failures, Flexachem.com

8

Proven Ways to Improve Steam Turbine Reliability

By Robert X. Perez and David W. Lawhon

Repairs versus Overhauls

Before we discuss steam turbine reliability, we must first talk about the difference between a repair and an overhaul. If a system, such as a steam turbine and its controls, can be restored to operating condition after a failure by the repair or replacement of one or more components, we say it is a repairable system. Since steam turbines are typically designed to allow minor field repairs, such as the replacement of seals, usually called packing, bearings, governor, etc., we can say they are repairable. However, after an extended time in operation, a steam turbine can no longer be repaired in the field; we say that is must be overhauled. An overhaul requires major work, such as replacing blading due to erosion, repairing rotor bearing journals, repairing leaking split-lines, etc.

By definition, we will define a steam turbine repair as the in-field replacement of a component or components in the field and an overhaul as the removal of the entire steam turbine from the field as a unit in order to restore it to factory specifications. Using these definitions, we will express the mean, or average, time between repairs as the MTBR and the mean, or average, time between overhauls as the MTBO.

Expected Lifetimes of Steam Turbines and Their Components

Table 8.1 contains a listing of expected lifetimes in hours and years of various components and mechanical systems. For comparison, the expected lifetimes of steam turbines are listed along with other types of process machinery. For example, the average lifetime of a centrifugal compressor is expected to be about 6.85 years, while the average lifetime of a steam turbine is somewhere between 5.71 and 7.42 years. The reader should note

Robert X. Perez (ed.) Maintenance, Reliability and Troubleshooting in Rotating Machinery, (181–188)

Table 8.1 Expected lifetimes of process machines and components.

Machine or component type	**Expected lifetime in hours**			**Expected lifetime in years**		
	Low	**Average**	**High**	**Low**	**Average**	**High**
Compressors, centrifugal	20,000	**60,000**	120,000	2.28	**6.85**	13.70
Compressor blades	400,000	**800,000**	1,500,000	45.66	**91.32**	171.23
Compressor vanes	500,000	**1,000,000**	2,000,000	57.08	**114.16**	228.31
Diaphragm couplings	125,000	**300,000**	600,000	14.27	**34.25**	68.49
Gas turb. comp. blades/vanes	10,000	**250,000**	300,000	1.14	**28.54**	34.25
Gas turb. blades/ vanes	10,000	**125,000**	160,000	1.14	**14.27**	18.26
Motors, AC	1,000	**100,000**	200,000	0.11	**11.42**	22.83
Pumps, centrifugal	1,000	**35,000**	125,000	0.11	**4.00**	14.27
Steam turbines (Barringer)	11,000	**65,000**	170,000	1.26	**7.42**	19.41
Steam turbines (NPRD & Davidson)	33,333	**50,000**	100,000	3.81	**5.71**	11.42
Steam turbine blades	400,000	**800,000**	1,500,000	45.66	**91.32**	171.23
Steam turbine vanes	500,000	**900,000**	1,800,000	57.08	**102.74**	205.48
Low pressure (<50 psi) steam turbine carbon ring packing (Reeves) (see note 3)	8760	**13140**	17520	1	**1.5**	2

(Continued)

Table 8.1 Expected lifetimes of process machines and components. (*Continued*)

Machine or component type	**Expected lifetime in hours**			**Expected lifetime in years**		
	Low	**Average**	**High**	**Low**	**Average**	**High**
Medium pressure (50 to 200 psi) steam turbine carbon ring packing (Reeves) (see note 3)	17520	**21900**	26280	2	**2.5**	3
High pressure (200 to 400 psi) steam turbine carbon ring packing (Reeves) (see note 3)	26280	**35040**	43800	3	**4**	5
Sleeve bearing	10,000	**50,000**	143,000	1.14	**5.71**	16.32
Roller bearings	9,000	**50,000**	125,000	1.03	**5.71**	14.27

Table Notes:

1. The basis for this table is Barringer's Weibull database: http://www.barringer1.com/wdbase.htm. The authors removed the Weibull shape factor data and only used the characteristic lifetimes to create this table.
2. These characteristic lifetimes are highly dependent on 1) steam quality, 2) installation quality, 3) installation design, 4) operating procedures, and 5) PM/PDM procedures.
3. High end carbon ring packing life readings assume TC has been applied to packing areas.
4. Source NPRD-95: Non-electronic Parts Reliability Data 1995, Reliability Analysis Center, Rome, NY
5. Davidson, John, 1988, *The Reliability of Mechanical Systems*, Mechanical Engineering Publications Limited for the Institution of Mechanical Engineers, London.

that this is the average of many different failure modes related to steam turbines. This should not be confused with the time between overhauls. From overhaul to overhaul there may be component failures that bring down the average lifetime of a steam turbine. (In the author's experience, steam turbines can routinely operate up to 10 to 12 years between major overhauls.)

Common Failure Modes

Between major overhauls there are four types of common failures experienced in steam turbines:

1. Steam packing leaks
2. Bearing failures, usually related to lubrication issues
3. Governor failures
4. Sticking trip and throttle valves

Steam Turbine Leaks

One striking feature of the Table is that expected lifetime of packing is significantly shorter than that of the expected lifetime of steam turbines. As seen in the "Expected Lifetime" table above, the average lifetime of packing can be greatly affected by the type of service that it is in and if tungsten carbine coatings are employed on bearing journals. Keep in mind that the average life (MTBR) of carbon ring packing is highly dependent on 1) steam quality, 2) installation quality, 3) installation design, 4) operating procedures, and 5) PM/PDM procedures. Break-in procedures should always be carefully followed to ensure a proper break-in after packing replacement.

Bearing and Lubrication Failures

Bearing failures are normally caused by lubrication contamination. To prevent these types of failures, you must install effective bearing housing seals, frequently inspect the packing for leaks, and test the lubrication oil for water content. Tattletale sight glasses are helpful in allowing you to stop contaminated or degrading oil.

Bearing failure rates can be significantly improved by utilizing a close loop lubrication system with a reservoir, coolers, oil coolers, and flow control piping.

Governor Failures and Sticking T&T Valves

When a governor or its trip and throttle valve begin to fail, you will notice irregular or periodic speed fluctuations, differences in regulation between cold and warm operation, or the nominal speed cannot be achieved. These problems may be related to either a governor, linkage, or trip and throttle valve issue. Before attempting a repair on any of these components, first check to see you have a reliable steam supply. If the steam supply is not an issue, then check for governor oil to ensure is the correct type and is not contaminated and that the linkages or trip and throttle valves are not sticking.

It is very common to have to replace or refurbish the hydraulic governor on a general purpose steam turbine. For this reason, it is highly recommended that a replacement governor be kept in warehouse stores.

Improvement Reliability by Design

To maximize steam turbine reliability, we must maximize the mean time between overhauls and maximize the mean time between repairs (MTBR) of steam turbine components. To this end operations and maintenance must strive to ensure their steam turbines are continuously supplied with high-quality steam. Operations and maintenance's secondary emphasis must be on maximizing the time between common failure modes mentioned above by optimizing design standards, operating procedures, repair standards, frequent inspections, and time-based maintenance.

Let's consider the effects that packing, coupling, and bearing lifetimes have on the overall MTBRs. Table 8.2 explores the theoretical reliability of various steam turbine design configurations. For example, in the first row is a steam turbine with carbon rings, a lubricated coupling, and oil sump lubrication. In all cases, we assume a steam path lifetime of 15 years. (The steam path lifetime is defined as the operating hours required to erode steam path components, i.e., rotating and stationary blading, to the degree that steam turbine performance is significantly degraded.) In the column at the far right, we calculate the average mean time between repairs for the entire unit. The equation for the MTBR is as follows:

$$\frac{1}{MTBR_{overall}} = \frac{1}{MTBF_{Seals}} + \frac{1}{MTBF_{Coupling}} + \frac{1}{MTBF_{Bearings}} + \frac{1}{MTBF_{Steam\ path}}$$

It is clear from Table 8.2 that for various design configurations the overall MTBRs vary from 1.11 years to 4.62 years. This exercise demonstrates that the average time between steam turbine outages is not controlled by the time between overhauls but by the time between component failures. We can therefore conclude that to maximize reliability of steam turbines is to

Table 8.2 Mean t/ime between repairs (in yrs.) for various design configurations.

Packing	MBTR1	Coupling	MTBR2	Bearing lubrication	MTBR3		MTBR4	Overall
Carbon rings	3	Lubricated	6	Oil Sump	3	Steam Path	15	1.11
Carbon rings	3	Dry	20	Oil Sump	3	Steam Path	15	1.28
Carbon rings	3	Dry	20	Closed Loop	20	Steam Path	15	2.00
Carbon rings with TC coating	5	Lubricated	6	Oil Sump	3	Steam Path	15	1.30
Carbon rings with TC coating	5	Dry	20	Oil Sump	3	Steam Path	15	1.54
Carbon rings with TC coating	5	Dry	20	Closed Loop	20	Steam Path	15	2.73
Labyrinth packing	20	Lubricated	6	Oil Sump	3	Steam Path	15	1.62
Labyrinth packing	20	Dry	20	Oil Sump	3	Steam Path	15	2.00
Labyrinth packing	20	Dry	20	Closed Loop	20	Steam Path	15	4.62

maximize the reliability of its weakest components, i.e., sealing systems, bearings, governor, and lubrication systems, etc.

The authors have found the following improvements have been proven to provide a significant improvement in reliability:

1. Apply tungsten carbon coatings at the rotor packing ring areas
2. Replace lubricated couplings with dry couplings
3. Install bearing housing seals to prevent oil sump contamination
4. In severe duty services, install a closed loop lubrication system

To summarize, we will reiterate the effects of steam turbine design standards and operating, repair, and PM/PDM practice on the average time between repairs. If we want to maximize turbine reliability, we must watch the details. No detail is too small when dealing with steam turbines. Both the operators and the mechanics must remain vigilant and communicate frequently in order to keep their steam turbines operating efficiently, reliably, and safely.

Acknowledgements

I would like to thank David Lawhon, Rotating Equipment Manager at the Motiva Refinery for his help with this chapter and Ron Reeves, Manager/Partner at Turbonetics Engineering & Services for providing the steam turbine reliability data and design improvement ideas discussed in this chapter.

9

General Purpose Steam Turbine Reliability Improvement Case Studies

By Abdulrahman Alkhowaiter

The goal of this chapter is to provide proven upgrade ideas to enable end users to improve general purpose steam turbine reliability. Outdated steam turbine designs are susceptible to problems, such as wear, galling, and fouling, that can result in frequent maintenance and unacceptable safety performance. Industry reliability leaders have shown that there are now many opportunities to improve the dependability of steam turbines. Here are some proven design upgrades and improvements that are discussed in this chapter:

- Governor valve upgrades
- Steam seal upgrades to dry gas seals

Robert X. Perez (ed.) Maintenance, Reliability and Troubleshooting in Rotating Machinery, (189–204)

- Fast start modifications
- Upgrades to electronic speed governor with pneumatic actuator and pneumatic trip system
- Lube oil system improvements
- Dry flexible metal coupling upgrades
- Baseplate improvements

The author will present case studies featuring successful design improvements that were made on problematic steam turbines. It is hoped that the variety of different design improvements covered in this chapter will provide users with a sense of what is possible with the latest steam turbine technology.

Governor Valve Packing Gland Leakage: Sealing & Reliability Improvements

The following design improvements were made to General Purpose Steam Turbines, by upgrading to the new governor valve integral design bushing-gland below. This changed the MTBF of the packing from 4 months to 4 years, whereas before, constant maintenance was required. A total of 20 Elliot & Westinghouse GP steam turbines were successfully modified to the new inlet valve gland design from 1996 to 2001.

1. The following Elliott parts were discarded: The Gland Follower, and the Lantern ring (See Figure 9.1). The two drawings below are for the Elliott conversion, with exact dimensions.
2. A new single piece guide bushing, which incorporates packing and a new gland follower, was designed and fabricated in our shop (Figure 9.2). This guide bushing is press fitted into the valve cover flange, which has been modified in step-4. Use a 0.0025 inch press fit and add grease before press-fitting. The guide bushing has three 13/64 inch holes, which are the 60 # intermediate steam leak-off holes. This substitutes the function of the original Lantern ring. Note: Do not undertake the final boring of the inlet valve stem hole (diameter: 0.504 inch) until the guide bushing has been press-fitted or shrink fitted into the valve cover (step 4). First screw

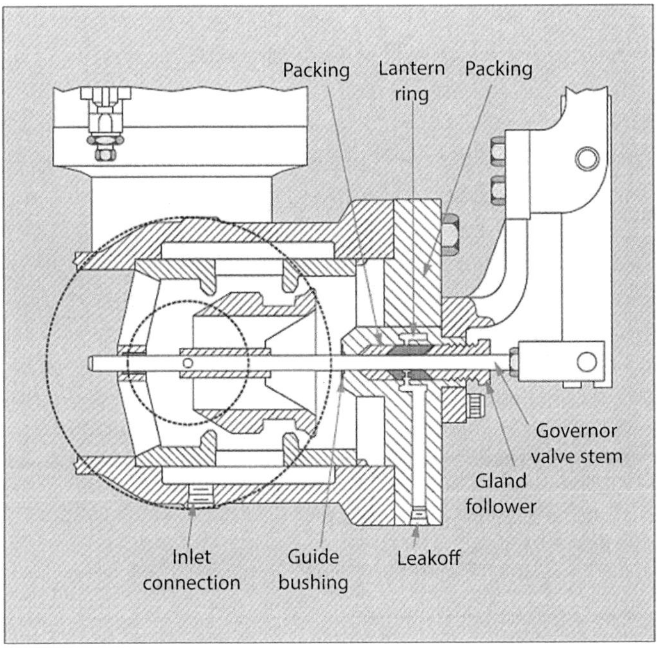

Figure 9.1 Excessive steam leakage was a constant reliability problem. This prompted us to discard the gland follower and the lantern ring and upgrade it with an integral bushing-lantern ring.

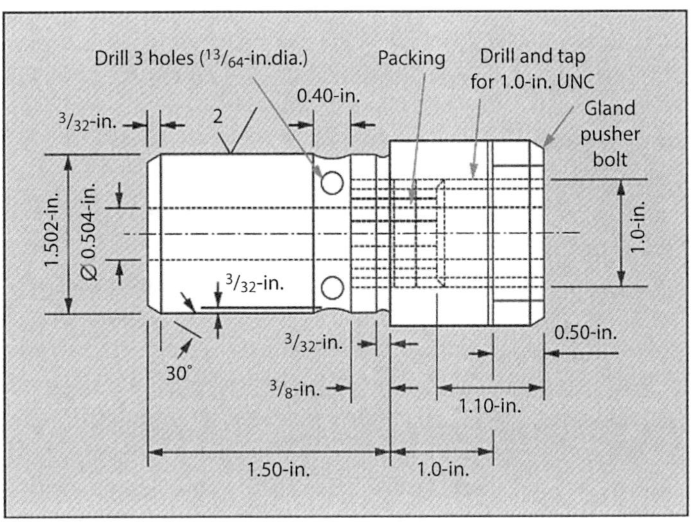

Figure 9.2 The outer gland was converted to an integral bushing-lantern ring.

in the gland pusher bolt, then do the final machining, to achieve perfect concentricity of all parts.

3. The new gland pusher is a short, fabricated bolt (1.0 inch UNC thread) which screws into the guide bushing. Only three to four pieces of standard steam packing are inserted, and then the gland pusher bolt is screwed in.
4. The governor valve cover flange (which the gland follower normally screws into) was modified: Place the valve cover flange on a lathe and bore out the original threaded hole. Bore out to a hole diameter of exactly 1.4995 inch (+ 0.0000, - 0.0005). Add a 1/16 radius to the hole inlet, to allow for smooth press fitting of the new bushing. Now, you can press fit the new throttle valve guide bushing into the machined valve cover. Be careful: once in service, there is 600 psig of steam pressure trying to force out the new bushing. Therefore, it is very critical to have an interference fit of 0.0025 inch between the bushing and valve cover. The material choice for new guide bushing was either 4140 or 4340 steels.

Steam Turbines Carbon Seals Upgrade to Mechanical Seals

The majority of GP steam turbines (see Figure 9.3) are currently sealed on the shaft with split carbon rings. In reality, the carbon rings do not really seal – they only control the leakage of steam to atmosphere. This results in the following:

- **Loss of steam**
- **Lower turbine efficiency**
- **Lower equipment (bearing) reliability due to lube oil contamination**
- **Frequent overhauls due to shaft wear at carbon seal area**

The solution to this sealing and reliability problem is to retrofit dry, non-contacting mechanical shaft seals to the turbines (see Figure 9.4). The author has retrofitted about 50 general purpose steam turbines since 2002, to the Wavy Face mechanical seals, which are made specifically for steam services up to 625 lb. inlet steam. The steam seal average MTBF has so far exceeded 6.0 years. Advantages realized after mechanical seals retrofit installation:

Figure 9.3 General purpose steam turbine.

Figure 9.4 Close-up of steam seal gland.

Typical Benefits of Dry Gas Seal in a 1500 HP Turbine

- A safer sealing system, reducing potential for injury from leaking steam.
- More reliable longer life expectancy, so far the seal installations at various Plants have exceeded 6 years MTBF.

- Reduced steam leakage by 98%, improving efficiency of the turbine.
- Eliminated the need for gland condenser, steam extraction and associated piping, reducing maintenance cost.
- Eliminated shaft wear caused by contact with carbon rings.
- Warm-up and run-in cycles are not required (carbon seal procedure).
- Avoided investment and maintenance of dehydration unit for the oil system.
- Reduced noise levels and corrosion to structures around steam turbine.
- The most reliable DGS seal face design in steam service on about 80 turbines installed at our facilities, has been Wavy Face design.

Table 9.1 summarizes the economics of upgrading from carbon rings to a dry gas seal. Notice the payback period for the upgrade project period was calculated to be only 12 months.

Table 9.1 Typical economic justification, 1500 HP turbine DGS installation.

Steam seal type	Carbon ring	GTS mechanical seal
Cost of Seals (Both INB/OUTB)	$1,000	$20,000
Initial Retrofit Cost	0	$5,000
Labor on seal repairs per annum	$2,160	0
Parts/Labor on repairs (bearings, etc.) per annum.	$3,000	0
Preventative maintenance costs per annum	$1,500	$1,500
Maintenance on auxiliary equipment per annum	$1,120	0
Steam losses per annum	$18,000	$200
Total Operating Costs per annum	**$26,780**	**$1,700**
Payback Period		**12 Months**

Modification of GP Turbines for Fast Start without Slow Rolling

The slow rolling of general purpose steam turbines in process plants causes a wasteful loss of steam energy and condensate. This leads to continuous component wear and frequent maintenance shutdowns. The following fast start, no slow roll procedure is recommended for general purpose, well-insulated, single-stage turbines of all sizes. The potential energy savings due to elimination of slow-rolling are, per 1000 HP turbine = $100,000 per year.

Critical Considerations during startup: The objective is a gradual temperature rise:

- Uniform growth of components
- Acceptable clearances
- Avoidance of thermal shaft bow
- Removal of any liquid condensate

How the GP Turbine Fast Startup Modification Works

- Install three steam traps (See Figure 9.5)
- Implement minor drain piping work as below

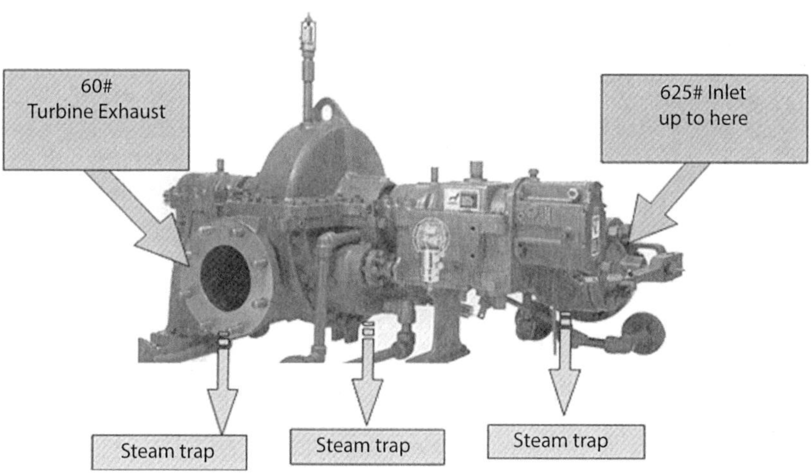

Figure 9.5 Steam trap configuration required for starting.

- Apply full thermal insulation to turbine casing
- Slow leakage of 60 lb. or 600 lb. steam through turbine, causes heating
- Modify startup sequence to allow instant start
- Issue letter to operations prohibiting slow-rolling

Dry Flexible Metal Coupling Upgrade with Split Spacer, for Short Coupled Turbines with Insufficient Length Coupling Spacers

The special design couplings (Figures 9.6 and 9.7) greatly reduce vibration problems arising from misalignment when existing shaft couplings have short or no spacers. The Coupling Corporation of America CCA couplings have a unique design because they are based on the principles of a diaphragm coupling in that they carry torque between an inner and outer diameter, but they differ from the standard diaphragm coupling design because they use multiple pieces of uniform thickness stainless steel instead of a single machined profile diaphragm.

The major benefit of this design is the very low spring rates, meaning that the amount of force taken to flex the coupling is extremely low.

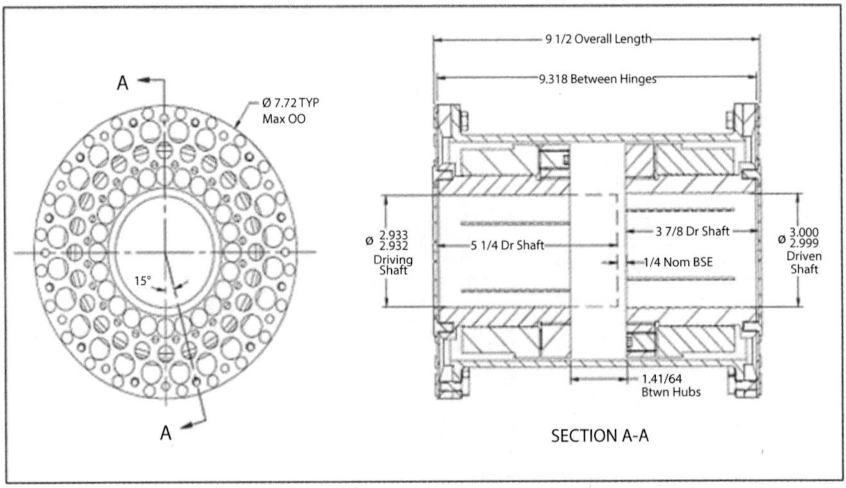

Figure 9.6 Split coupling spacer design, installed on 2000 HP, pump-turbine-gearbox set.

Figure 9.7 Field photo of dry flexible metal coupling upgrade with split spacer.

Typically, the spring rates are at least 10-15 times lower than other types of couplings, including gear, disc, diaphragm, and rubber. A quick example, for a 0.010 inch axial movement, standard diaphragm or disc coupling might require 20 lbs. of force, which is a spring rate of 2000 lbs. per inch. With CCA couplings, a 0.010-inch axial movement would require 1.5 lbs., which is a spring rate of 150 lbs. per inch.

The major benefit of the lower spring rates is that the force required to flex the coupling must come from somewhere, and it comes from either the shaft bending, or the bearings being stressed. Therefore, in the long run, the coupling can be a lifesaver for both coupled machines. The other benefit is greatly reduced vibration transmission through the coupling.

Keyless Hubs: The other improved product is the CCA Anderson Clamp Hub, which is a keyless hub that clamps on to a straight, tapered, or keyed shaft. The hub is actuated by a set of pushing screws, which push a collar up a taper, forcing the body of the hub to clamp on the shaft. This device is different from other taper lock type devices, and it will carry as much torque as a standard keyed hub. It is much cleaner, quicker, and easier than using a hydraulic hub. Also keep in mind that the Anderson Clamp Hub and our couplings can be used independently, i.e., the Clamp Hub can be used with other couplings, and the couplings can be used with other hubs.

Figure 9.8 Upgraded lube oil system.

General Purpose Lube Oil System Upgrade for Self-Contained Bearing Housings to Eliminate Overheating & Bearing Failures

On larger general purpose steam turbines, bearing overheating can occur due to limited cooling in summer. One solution is to fabricate or purchase a small lube oil circulating system (see Figure 9.8) with heat exchanger for turbines of 1000 HP rating and above.

Governor and Trip System Upgrade from Hydraulic to Electronic-Pneumatic

The original hydraulic-governor & overspeed trip system (see in Figure 9.9) was highly unreliable. This led us to convert to an electronic-pneumatic system.

Governor Requirements:

1. Governor actuators shall be of the pneumatic piston type or Electric Actuators, acting directly on the throttle valve, with no linkages. The governor shall be installed at a maximum distance of 4.0 meter from the turbine.

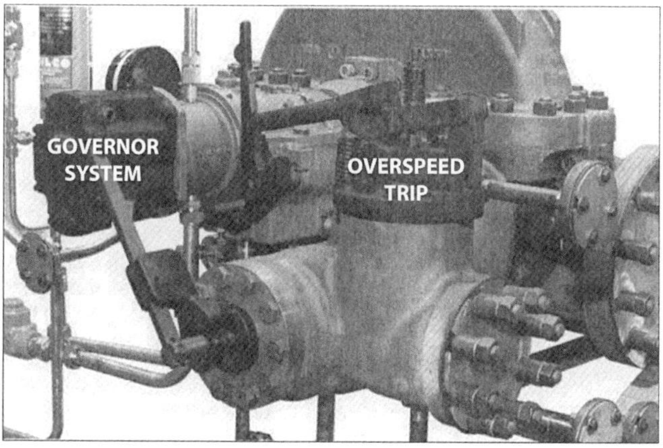

Figure 9.9 The original hydraulic-governor & overspeed trip system.

2. Electronic governors for general-purpose steam turbines shall be of the simplex type. Loss of electrical power to the governor or emergency trip device, shall lead to an automatic trip. Loss of governor signal shall close the governor valve.
3. Governors shall be suitable for outdoor installation and shall be provided with a sunshade. The maximum distance from a turbine to a governor will depend on the type of instrumentation. For passive pickups and transducers, the distance shall be limited to 75% of the maximum distance providing satisfactory operation of the control system. Longer distances shall require active pickups and transducers with power supply or modulated current transducers.

Electronic Governor with Pneumatic Actuator & Pneumatic Trip System

The new system has pneumatic actuators as trip valve actuators (Figure 9.9), and direct action solenoid trip valves for air shutoff, and simplified speed probe installation. No hydraulics, no mechanical trips.

Governor and Trip System Requirements

1. Overspeed Trip System: A single valve with combined Trip and Throttle functions is not acceptable. A separate air operated globe type trip valve with spring return shall be supplied and designed for steam service with fail-closed design. The trip valve trim shall be of hardened stainless steel.
2. Mechanical trip systems and linkages to the turbine are not acceptable. The valve shall trip fully closed within 0.75 seconds maximum. Trip shall be through a minimum of two solenoid-operated pneumatic-valves set at fail-safe mode, with one out of two trip actions. Solenoid valve action shall unload air from trip valve cylinder. Solenoid valves to have H-rated coils and Viton or Teflon soft seating. The trip valve pneumatic actuator shall be of the piston-cylinder type, as seen in Figure 9.10, and shall be equipped with a separate hand lever with red color handle, for manual trip.

Figure 9.10 Upgraded elliott trip system with pneumatic actuator.

Overview of All-Electronic Trip and Overspeed Protection System

The traditional design of general purpose steam turbines incorporates mechanical systems having many moving parts that are susceptible to wear, galling, and fouling, which can result in frequent maintenance and unacceptable safety performance. Modern overspeed trip protection systems have been greatly improved, which has led to the rapid obsolescence and discontinuation of mechanical overspeed trip devices and older electronic speed protection devices. Upgrading to a modern electronic overspeed trip device safeguards the equipment and personnel working around the equipment from dangerous overspeed events. The new electronic overspeed trip system selected (see Figures 9.11 and 9.12) a) has fewer wearing parts than the old mechanical systems, b) improves communication with plant operations, and c) allows turbine operators to accurately monitor speed with ease. It was also believed that the new system would dramatically reduce spurious turbine trip events.

Outboard Bearing Improved Flex Foot: Higher Turbine Reliability & Lower Vibration

Steam Turbine Rear Bearing Flexible Feet Stiffness: The general purpose 2400 HP turbine seen in Figure 9.13 has been redesigned with

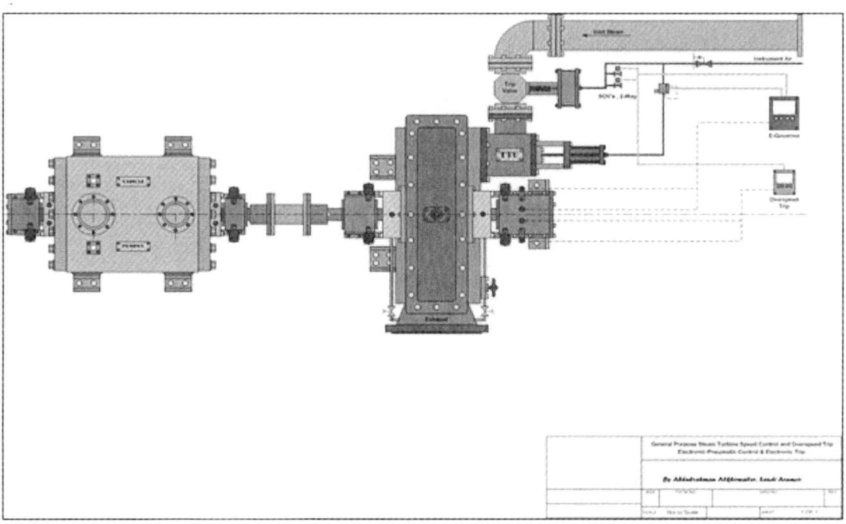

Figure 9.11 Overview of new overspeed protection system.

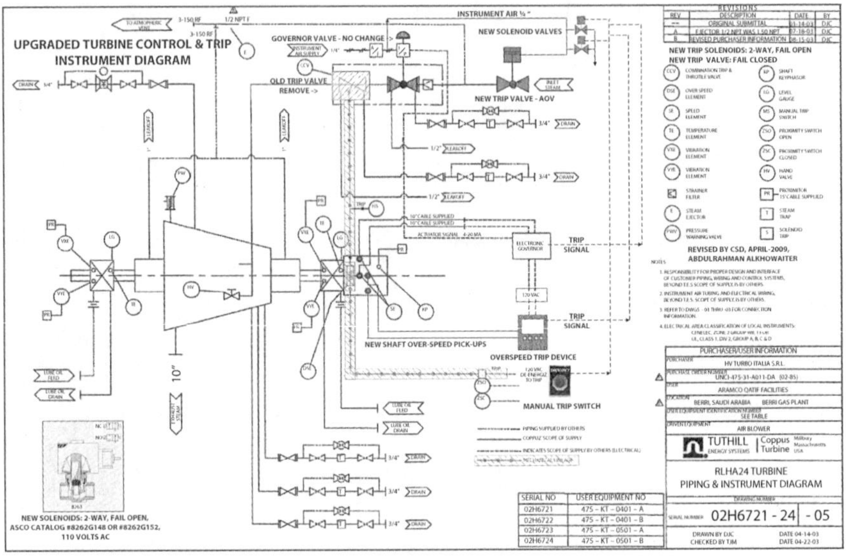

Figure 9.12 Schematic of new overspeed protection system.

new outboard bearings flex plate. The OEM support feet plates are too flexible; most are only 0.25 to 0.40 inch thick and do not provide sufficient rigidity against inlet steam pipe forces. This causes a tilting of the OB bearing housing and casing leading to high misalignment at

Figure 9.13 New heavy duty outboard bearing housing support flex foot.

coupling and long-term vibration problems. It also allows casing twist which introduces bearing misalignment. The solution is to retrofit a heavier section outboard bearing Flex-Plate to resist the misalignment of turbine from piping loads as seen in Figure 9.13.

Retrofit by fabricating a 5/8 inch to 3/4 inch thickness mild steel plate instead of original plate. This thickness still provides all the axial flexibility required for axial thermal growth of the turbine casing. The improved plate thickness selection depends greatly upon the height of plate, as it must allow an axial thermal growth of casing with 3 mm maximum growth.

Results

The result of implementing the design improvements described in this chapter was an average increase in the mean-time-between-steam turbine failures (MTBF) from 4 months to 4 years. These steam turbines are no longer considered to be maintenance problems. A total of 20 General Purpose Steam Turbines were successfully modified to the new inlet valve gland design from 1996 to 2001.

Part III
MAINTENANCE BEST PRACTICES

10

Rotating Machinery Repair Best Practices

By Robert X. Perez

World-Class Reliability Performance Should be the Goal of Every Repair Facility

Figure 10.1 Cutting corners during a repair might initially save time and money but will usually result in a sub-par repair that could end up affecting a machine's long-term reliability.

A maintenance best practice is any repair procedure or specification proven to ensure the plant asset will perform safely and reliably at the lowest overall cost of ownership (see Figure 10.1). World-class maintenance organizations accumulate a collection of best practices to cover a wide range of machinery activities so that they can consistently perform topnotch maintenance. However, there are times during maintenance activities when we are tempted to cut corners in the interest of time and money, or because

Robert X. Perez (ed.) Maintenance, Reliability and Troubleshooting in Rotating Machinery, (207–226)

of lack of resources, even though we know that we should always follow maintenance best practices. Allowing plant personnel to cut corners on a regular basis will result in subpar reliability performance. Over time, an organization can fall into a sub-par culture instead of doing the right thing the first time. Experience has taught us that best practices have been proven to provide the optimum level of safe and reliable service.

Cutting Corners = Unreliability

Here are a few maintenance examples of cutting corners around plant machinery:

- Not balancing or check balancing a rotor before it's installed. Rotor balancing is key to minimizing machine vibration and maximizing reliability.
- Not performing a high-quality alignment. A good machine alignment is also important to minimizing machine vibration and maximizing reliability (see Figures 10.2 and 10.3).
- Not checking critical fits and clearances. Proper fits and clearances are required to ensure reliable machine operation. Here are two examples of cutting corners related to fit that is out of tolerance components: a) sleeving a bearing housing to restore an ID fit, which is not an accepted best practice, or b) coating a shaft to restore an OD dimension, which will lead to more problems if not done correctly.

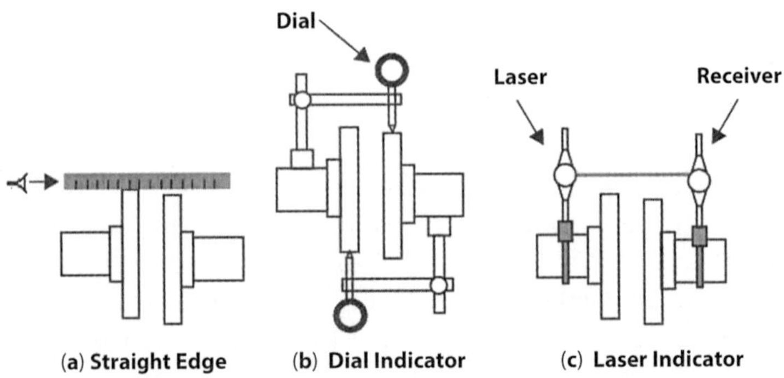

Figure 10.2 The three most common alignment methods: (a) straight edge method, (b) dial indicator method, and (c) laser indicator method.

- Using parts that are out of tolerance, i.e., has too much run-out, is out of round, or doesn't have the right surface finish, can lead to a shortened operating life.

Maintenance personnel understand that their primary responsibilities are to keep the equipment running and minimizing downtime in the event of an unpredicted failure or breakdown. Although it's important to make repairs and perform the needed maintenance quickly, it is equally important to perform these repairs and duties correctly and to their completion. We've all experienced a time where a critical machine is down, and production or operations is pushing you to make the repairs as quickly as possible so the machine can be put back into service [1].

This can create a situation where it may be tempting to cut corners, skip steps, and fail to ensure the repair procedures were done properly and completely. There are two possible outcomes that could result in cutting corners: 1) potential safety concerns, 2) reduced MTBF (mean time between failures) and shortened overall life span of the equipment.

The Importance of Alignment

Shaft alignment is the positioning of the rotational centers of two or more shafts such that they are co-linear when both machines are under normal operating conditions. There are two components of misalignment: a) parallel offset, or misalignment, and b) angular misalignment, or angularity (see Tables 10.1a and 10.1b). With parallel misalignment, the centerlines of both shafts are parallel, but they are offset. With angular misalignment, the shafts are at an angle to each other.

Angular misalignment can be a confusing concept to those new to the field of alignment. Compared to angular misalignment, parallel misalignment is a pretty straightforward idea, which can be measured within a few thousandths of an inch with a good straight edge. Basically stated, angularity is the gap difference across two coupling faces-normally top to bottom, or side to side. The three common alignment methods used today, listed in order of worst to best, are a) straight edge method, the b) dial indicator method, and the c) laser indicator method (see Figure 10.2).

When field vibration data from a misaligned machine is analyzed with a vibration analyzer, typically 1x, 2x, and sometimes 3x components are found in the vibration spectrums. These components may appear in spectra when taken using accelerometers oriented in the radial and axial directions (see Figure 10.3).

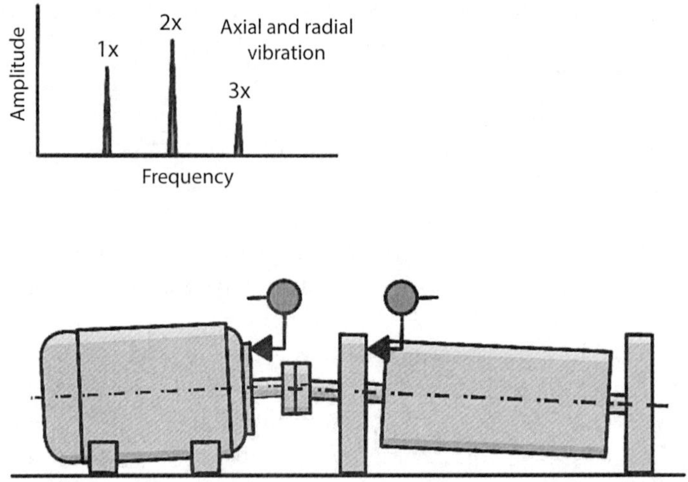

Figure 10.3 Misalignment shows up as 1x, 2x, and sometimes 3x components in the vibration spectrums.

Alignment Tolerances [2]

The desired target of alignment is line-on-line or two shafts rotating on the same line. These alignment rules of thumb serve as general guides for general machinery:

Alignment target: 0.0005 in/in of space between flexing elements, i.e., gear teeth engagement or coupling spacer length.

DBSE: Gear type +/- 0.030 in of the distance specified; Flexing elements +/- 0.010 in.

Alternative Alignment Guidelines

Shaft to shaft alignment should be carried out within the recommended tolerances in Tables 10.1a and 10.1b, unless a more precise specification is required by the manufacturer or the owner.

Tables 10.1a and 10.1b, Shaft Alignment Tolerances. Tolerances presented here are the maximum allowable from zero offset and zero angularity specifications. Any intentional targeted offset or angularity alignment values must be added to these recommendations. (E indicates English units and M indicates metric units).

Table 10.1a For couplings with short or no spacers.

Vertical & horizontal parallel Offset 				**Angularity (gap)** 			
				Coupling diameter			
RPM		**OFFSET**		**10 in.(250mm) or less**		**10 in.(250) mm) or greater**	
E	M	E	M	E	M	E	M
600	600	.005"	0.13 mm	.005"	0.13mm	.0005"/ in	0.05mm/ 100mm
900	750	.003"	0.08 mm	.003"	0.08 mm	.0003"/ in	0.03mm/ 100mm
1200	1000	.0025"	0.06 mm	.0025"	0.06mm	.00025"/ in	0.03mm/ 100mm
1800	1500	.002"	0.05 mm	.002"	0.05mm	.0002"/ in	0.02mm/ 100mm
3600	3000	.001"	0.03 mm	.001"	0.03mm	.0001"/ in	0.01mm/ 100mm
7200	6000	.0005"	0.01 mm	.001"	0.03mm	.0001"in	0.01mm/ 100mm

Alignment Calculation Example

A coupling with 6" coupling hubs and an 8" spacer in a 3600 rpm service should be aligned with a vertical or horizontal offset of less than 0.0003 x 8"=0.0024" (0.061mm) and an angularity of less than 0.001" (0.03 mm).

Rotor Balance

Rotor unbalance, or imbalance, is the uneven distribution of mass around an axis of rotation. A rotating mass, or rotor, is said to be out of balance

Table 10.1b For couplings with spacers.

Vertical & horizontal offset per inch (25.4 mm) of spacer length 				**Angularity (gap)** 			
				Coupling diameter			
RPM		**OFFSET**		**10 in.(250mm) or less**		**10 in.(250) mm) or greater**	
E	M	E	M	E	M	E	M
600	600	.0018"	0.457 mm	.005"	0.13 mm	.0005"/ in	0.05mm/ 100mm
900	750	.0012"	0.0305 mm	.003"	0.08 mm	.0003"/ in	0.03mm/ 100mm
1200	1000	.0009"	0.0229 mm	.0025"	0.06 mm	.00025"/ in	0.03mm/ 100mm
1800	1500	.0006"	0.0152 mm	.002"	0.05 mm	.0002"/ in	0.02mm/ 100mm
3600	3000	.0003"	0.0076 mm	.001"	0.03 mm	.0001"/ in	0.01mm/ 100mm
7200	6000	.00015"	0.0038 mm	.001"	0.03 mm	.0001"in	0.01mm/ 100mm

when its center of mass is out of alignment with the center of rotation. The units of unbalance are mass times radius. The weight of balance correction multiplied by the applied radius is the unbalance correction. Here are the most common units for expressing unbalance:

Imperial Units

- gram inches (g in)
- ounce inches (oz in)

Metric Units

- gram millimeters (g mm)
- gram centimeters (g cm)

For example, one ounce inch of unbalance would be a heavy spot of one ounce located at a radius of one inch from the rotating centerline.

There are two main categories of unbalance: static unbalance and dynamic unbalance.

Static Unbalance

Static unbalance is the simplest form of unbalance. It is equivalent to a single heavy spot in the rotor and can be corrected by adding balance weights to a single plane balance.

Dynamic Unbalance

Complex rotors typically have an infinite number of imbalances distributed randomly along its axis of rotation, which can only be assessed and corrected when the component is rotating. This kind of random imbalance distribution is referred to as dynamic unbalance. Dynamic unbalance in a rigid rotor requires balance correction in 2 planes. (A rotor that operates at a rotational speed below 70% of its critical speed is considered to be a rigid rotor. The critical speed is the speed at which the natural frequency of the rotor is excited. Ninety percent or more of industrial rotors are rigid.) Flexible rotors often required balance corrections at more than two planes.

Balancing

Balancing is the process of adding or removing weight from the rotor to align the effective mass center with the axis of rotation. The ultimate goal of balancing is to add the proper correction weights, at the right locations, to offset the net imbalance on a troublesome rotor in order to minimize rotor shaking forces.

Rotor unbalance generates forces that cause a rotor and support structure to vibrate at a frequency equal to running speed, commonly called the

1x frequency. Vibration associated with imbalance is the most common type of vibration found around industrial machines, which is why great time and effort is expended to balance rotors to accepted standards. In the case of a flexible rotor, we hope to ensure that vibration levels remain below an acceptable magnitude of rotor deflection at all speeds, up to the maximum service speed. Most rotors are balanced before machine assembly because of the limited access to the rotor in the field. Most balancing standards, such as ISO and API standards, classify rotor in accordance with their balancing requirements and establish methods of assessment of residual unbalance.

When analyzed with a spectral analyzer, imbalance tends to show up as a predominant 1x peak in the vibration spectra. These high 1X amplitudes show up predominantly in the radial direction, i.e., normal to the axis of rotating. X-Y radial 1x components should be approximately equal. If the horizontal and vertical 1x vibration differ significantly in amplitude, consider the possibility of some type of looseness or a structural resonance.

Common Causes of Rotor Unbalance [3]

Manufacturing: Many causes are listed as contributing to an unbalance condition, including material problems such as density, porosity, voids and blowholes; fabrication problems such as misshapen castings, eccentric machining and poor assembly; and distortion problems such as rotational stresses, aerodynamics and temperature changes. Many of these occur during manufacture, others during the operational life of the machine. While some corrections for eccentricity can be counteracted by balancing, it is a compromise. Dynamic balancing should not be a substitute for poor machining or other compromise manufacturing practices. In the manufacturing process, if proper care is taken to ensure that castings are sound and machining is concentric, then it follows that the two axes will coincide, and the assembled rotor will be in a state of balance (Lyons, 1998) [4].

Assembly: There are many reasons why unbalance occurs when a rotor is being fabricated. Principle among these is the possibility of tolerances. When a well-balanced shaft and a well-balanced rotor are united, the necessary assembly tolerances can permit radial displacement, which will produce an out of balance condition. The addition of keys and keyways adds to the problem. Thus, when a unit is assembled and the permanent key is added, unbalance will often be the result. The modern balancing tolerances specified by ISO, API, ANSI and others make it imperative that rotor assembly conventions be followed. Failure to do so will mean that the low-level balance tolerances specified in these standards will be impossible to achieve (Lyons, 1998).

Imbalances Associated with Installed Machines: When a rotor has been in service for some time, various other factors can contribute to the balance condition. These include corrosion, wear, distortion, and deposit build up. Deposits can also break off unevenly, which can lead to severe unbalance. This particularly applies to fans, blowers, compressors and other rotating devices handling process variables. Routine inspection and cleaning can minimize the effect, but eventually the machines will have to be removed from service for balancing. Large unbalances will of course require large weight corrections and unless care is taken, this can have a detrimental effect on the integrity of the rotor. Furthermore, concentrating a weight adjustment (whether adding or taking away) at a given point can weaken the rotor.

Other Causes: Another cause of unbalance that is not readily apparent is the difference between types of rotors. There are two distinct types of rotors, rigid and flexible. A rotor operating over or just under (say, within 20-25%) its first critical speed (i.e., the natural frequency RPM speed at which resonance occurs) is a flexible rotor. If it is operating below that speed, it is considered rigid. A rigid rotor maintains a straight centerline at its normal operating speeds and can be shop-balanced or *in situ* balanced in two planes. A flexible rotor, by definition, is running above its natural frequency and, accordingly, the geometric centerline of the shaft will "flexibly" follow the mass centers of the various major components.

A rigid rotor can be balanced at the two end planes and will stay in balance when in service. A flexible rotor will require multi-plane balancing. If a rotor is balanced on a low speed balancing machine assuming it is rigid and then in service becomes flexible, then unbalance, and thus high vibration, will be the result. Typical machines, which fit this category, include steam and gas turbines, multistage centrifugal pumps, compressors and paper rolls.

Balancing Grades

The two primary factors that determine permissible unbalance (also called the balancing tolerance) are the *mass of the rotating part* and the *maximum operational speed*. Calculating a balancing tolerance based on these parameters is relatively simple. However, these calculated tolerances apply to the journal planes of a rotor and must be converted for use at the actual rotor correction planes.

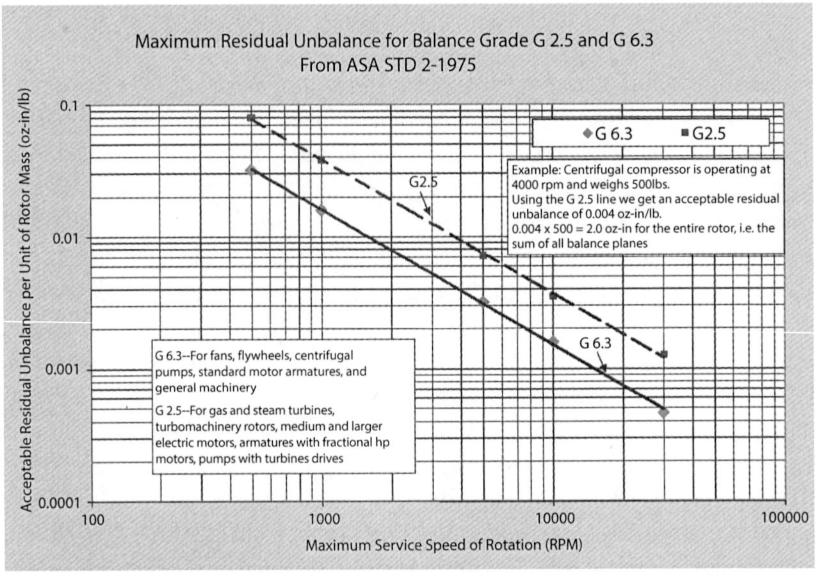

Figure 10.4 Maximum residual unbalance for industrial rotors (English Units).

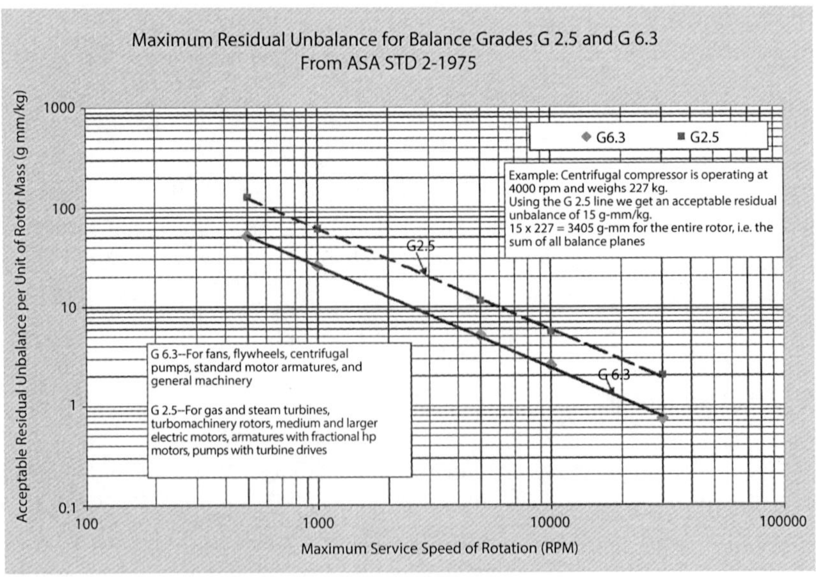

Figure 10.5 Maximum residual unbalance for industrial rotors (Metric Units).

Figures 10.4 and 10.5 show how to determine the acceptable amount of residual imbalance for G6.3 and G2.5 classified rotors. Grade G 6.3 should be used for fans, flywheels, centrifugal pumps, standard motor armatures, and general machinery; and Grade 2.5 should be used for gas and steam turbines, turbomachinery rotors, medium and larger electric motors. armatures with fractional hp motors, and pumps with turbine drives.

The Importance of Fit, Clearance & Tolerance

Let's now discuss fit, clearance, and tolerance standards and why they are an essential part of a site's rotating machinery repair best practices. Proper fits and clearances are essential elements of machinery reliability. Along with fit and clearance standards, tolerance standards are required to determine when a part is no longer serviceable and should be replaced. Let's briefly discuss what these key of machinery parameters are and why these standards should be utilized during every machine repair.

Fits, Clearances and Tolerances [5]

Rotating equipment is designed so that internal parts have either a) a gap between them, so that there is no contact between them, or b) they are firmly in contact with one another, so that there is no relative motion. The gap between two moving parts is called clearance, which is determined by the size difference between two moving parts. The dimensional difference between two non-moving parts is called fit. Fits can range from loose, i.e., slip fits, to tight, i.e., interference fits.

Tolerance

Because of gradual cutting tool wear and minute changes in the machine tool internals due to temperature changes and wear/movement of internal parts, machined items can not all be made perfectly to the same dimension. To ensure that machine parts will function properly, a dimensional tolerance is specified, which is defined as the acceptable difference between the maximum and minimum size of a mechanical part. It is permitted to make the part to within a range of sizes, which is called the tolerance on the dimension. Users should always ensure that internal parts have the proper clearances and fits that fall within the tolerances specified by the manufacturer.

Clearance

Examples of critical clearances are impeller wear rings that tend to gall. Galling is a form of wear caused by adhesion between sliding surfaces, whenever contacting surfaces are forced together. Galling accelerates whenever two metals are in close running clearance and develop a burr either due to abrasives in the pump media or rubbing. Stainless steels are examples of wear ring metallurgy that are particularly susceptible to galling. Typically, wear rings are designed with a hardness difference between the two rings to prevent galling.

When using wear ring metallurgical combinations that are susceptible to galling, add additional clearance to prevent contact between wear rings and ensure adequate preheating to ensure proper internal clearances are present. For pumping temperatures of 500°F and over, add 0.010″ to the wear ring clearance (see Table 10.2 below). Also, whenever galling-prone wear ring materials such as stainless steel are used, 0.005″ should be added to the clearance.

The type of fit varies. In an interference fit (see Figure 10.6), the parts are purposely made to be forced together. This type of fit can be further described as heavy through to light interference, whereas a clearance fit is for parts made to have a space between them. This fit can be further described as "tight" all the way to "loose." Between these two fits is the transition fit where interference may or may not occur. The amount of interference or clearance is achieved by specifying the tolerance range for the part's possible sizes.

Table 10.2 Recommended wear ring clearances.

Wear ring diameter	Diametrical clearance under 500 °F	Diametrical clearance under 500 °F
3.5"- 5"	0.016"	0.018"
5" – 6"	0.017"	0.019"
6" – 7"	0.018"	0.020"
7" – 8"	0.019"	0.021"
8" – 9"	0.020"	0.022"
9" – 10"	0.021"	0.023"
10" – 11"	0.022"	0.024"
11" and over	0.023"	0.025"

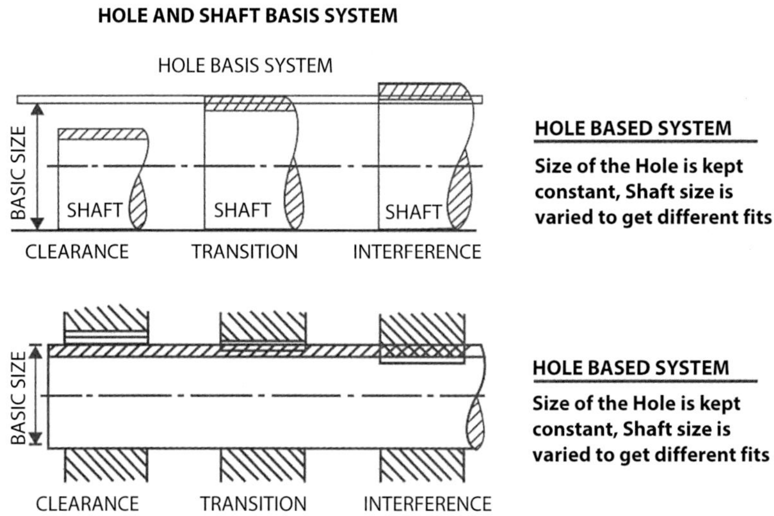

Figure 10.6 The classification of different types of fits.

Coupling Hub Fits

Couplings are common machinery components where interference fits are critical to reliability operation. Both keyed and keyless fits are used to secure coupling hubs to drive shafts.

Keyed Interference Fits [6]

Keyed interference fit couplings are commonly used for applications up to a few thousand horsepower/kW and speeds up to or slightly above 10,000 rpm. The interference fit standard for most couplings made of medium-carbon steel are 0.00050-0.00075 in/in (mm/mm) interference up to 1800 rpm and 0.00075-0.00100 in/in (mm/mm) over 1800 rpm. Table 10.3 lists coupling clearance fits used for common NEMA-frame-size motor shafts. The purpose of the interference fit with a keyed shaft is to axially locate the coupling hub and resist forces associated with unbalance and misalignment.

Keyless Interference Fits

For high-horsepower/kW and high-speed applications, ANSI/AGMA 9003-B08 or equivalent fits (straight and tapered) are commonly used. The interference for keyless fits needs to be adequate to withstand expected

Table 10.3 Coupling clearance fits for common NEMA-frame-size shafts [6].

Nominal shaft size	Decimal shaft size	Bore minimum	Bore maximum
5/8	0.625	0.6250	0.6260
7/8	0.875	0.8750	0.8760
1 - 1/8	1.125	1.1250	1.1260
1 - 3/8	1.357	1.3750	1.3760
1 - 5/8	1.325	1.6250	1.6260
1 - 7/8	1.875	1.8750	1.8760
2 - 1/8	2.125	2.1250	2.1265
2 - 3/8	2.375	2.3750	2.3765
2 - 7/8	2.875	2.8750	2.8765
3 - 3/8	3.375	3.3750	3.3765

normal and transient loads. Common keyless interference fits range from 0.0015 in/in (mm/mm) to 0.0020 in/in (mm/mm). The Brinell hardness of the hub material is a significant factor in keyless coupling fit. Typical fits for various Brinell hardness (BH) steels are: 0.00175 in/in (mm/mm) for 250 BH, 0.0025 in/in (mm/mm) for 300 BH and 0.0030 in/in (mm/mm) for 330 BH. Table 10.4 lists the interference fits used for common NEMA-frame-size motor shafts.

Effects of Excessive Looseness

An improper fit between component parts is generally characterized by vibration spectra with a string of rotating frequency harmonics, i.e., spectral peaks at 1x, 2x, 3x, 4x, 5x, 6x, etc. (Sometimes 0.5X, 1.5X, 2.5X components can appear in the spectrum.) Typically, the 1x component will be the largest and then each higher multiple will appear progressively smaller and smaller. NOTE: The amplitudes of these harmonics may be sporadic and can change drastically with load and temperature fluctuations.

Table 10.4 Coupling interference fits for common NEMA-frame-size shafts [6].

Nominal shaft size	Decimal shaft size	Bore minimum	Bore maximum
5/8	0.625	0.6240	0.6245
7/8	0.875	0.8740	0.8745
1 - 1/8	1.125	1.1240	1.1245
1 - 3/8	1.357	1.3740	1.3745
1 - 5/8	1.325	1.6230	1.6240
1 - 7/8	1.875	1.8730	1.8740
2 - 1/8	2.125	2.1230	2.1240
2 - 3/8	2.375	2.3730	2.3740
2 - 7/8	2.875	2.8730	2.8740
3 - 3/8	3.375	3.3720	3.3735

Rotating Element Looseness

The presence of loose rotating elements is often characterized by a spectrum containing multiple frequencies related to the rotating frequency of the rotor. Typically, when looseness is present, you can detect harmonics (1x, 2x, 3x, etc.), subharmonics (0.5x) and half harmonics (1.5x, 2.5x, 3.5x, etc.) of the fundamental rotating frequency (see Figure 10.7). These components are generally more predominant in the radial directions of measurement (horizontal and vertical) than in the axial direction.

As the wear increases, the frequency spectrum changes, allowing a classification of the severity of the looseness into the following four groups:

- Incipient looseness, which is characterized by presence of the first four or five harmonics of the rotating speed, showing less amplitude the higher harmonics.
- Potentially serious looseness, which is characterized by an increase of the amplitude of the first harmonic of the rotating speed and appearance of very low amplitude half harmonics.
- Serious looseness, which is characterized by an increase of the amplitude of the harmonics and half harmonics of the rotating speed.

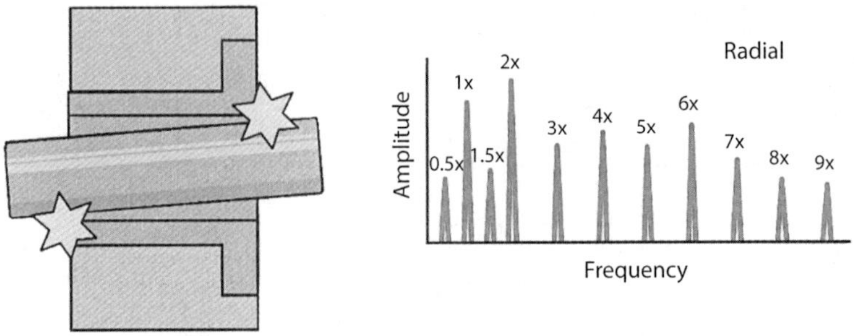

Figure 10.7 Typical spectrum of rotating element looseness.

- Severe looseness, which is characterized by the presence of 1) half harmonics, harmonics, and subharmonics of the rotating frequency of the rotor, and 2) a broad continuous energy band in the spectrum.

Effects of Internal Looseness

Be aware that internal friction between a shaft and shrunk-on elements, such as impellers, have been known to be a cause of whirl instability in built-up rotors since the early 1900s. This internal damping tends to make the rotor whirl at shaft speeds greater than a critical speed, the whirl speed usually being equal to the critical speed. Therefore, the fit of shrunk on rotor elements is critical to turbomachinery reliability.

Another potential problem caused by internal looseness is fretting. Fretting is the wear and oxidation due to repetitive sliding between two steel surfaces. The amplitude of the relative sliding motion is often in the order from micrometers to millimeters but can be as low as 3 nanometers. The contact movement causes mechanical wear and material transfer at the interface between the components, often followed by oxidation of both the metallic debris and the freshly exposed metallic surfaces. Because the oxidized debris is usually much harder than the surfaces from which it came, it often acts as an abrasive agent that increases the rate of fretting. Fretting can occur between mating rotor components that move relative to one another or between rolling elements and raceways. When signs of fretting are seen you should suspect improper shaft and bearing housing fits or relative movement between rotor elements.

Structural Looseness

Structural Looseness is another type of looseness that can be encountered around machinery whenever a machine or its associated piping is either 1) not properly bolted down, or 2) lacks adequate support stiffness. Excessive vibration levels related to structural looseness are typically highly directional, with the highest amplitudes oriented in the direction of lowest stiffness. Vibration that is predominantly 1x and highly directional tends to be related to structural looseness or a lack of stiffness.

If you suspect a structural looseness issue is present, first retorque all hold down bolts to ensure they are properly tightened. If hold down bolts are not properly designed or maintained, they will eventually fail, causing high vibration levels and transferring higher shaking loads to adjacent bolts. Mechanics must ensure hold down bolts are periodically inspected and retorqued to ensure their integrity.

If vibration is still high after torquing all hold down bolts, then the problem might be caused by a structural resonance or a general lack of support stiffness. An operating deflection shape (ODS) analysis could help identify the sources of the problem by showing you how the machine and support structure is vibrating in space. Another method of analyzing structural vibration issues is called motion amplification (MA), which is a video-processing method that detects subtle motion and amplifies it so that it is visible to the naked eye. The process involves the use of a proprietary high definition, high dynamic range video camera. MA can be used to identify vibration issues such as misalignment, imbalance, and looseness and can be particularly useful tool for analyzing and diagnosing structural defects and design issues.

We have briefly reviewed the most common assembly and installation problems found around rotating machinery: imbalance, alignment, and looseness. Table 10.5 lists the commonly encountered field vibration issues, along with how they can be identified using spectral analysis methods:

As Found and As Left Measurements

I recommend that you measure and record "before" (as found before the repair) and "after" (as left after the repair) component fits and clearances during every repair. This information will help future mechanics better understand how these fits and clearance are wearing or eroding during the machine's lifetime. Machine designers can also use this information to improve on their design to maximize operating life.

Table 10.5 Common assembly and installation problems and their associated vibrational characteristics.

Problem	Predominant frequencies	Notes
Imbalance	Indications of imbalance show up as a predominant 1x peak in the vibration spectrum.	Imbalance will be predominant in the radial directions. X-Y radial 1x components should be approximately equal.
Misalignment	Indications of misalignment can show up as 1x, 2x, and sometimes 3x components in the vibration spectrums.	These components may appear in spectra taken using accelerometers oriented in both the radial and axial directions.
Looseness (rotating)	Typically, when looseness is present, you can detect harmonics (1x, 2x, 3x, etc.), subharmonics (0.5x) and half harmonics (1.5x, 2.5x, 3.5x, etc.) of the fundamental rotating frequency.	As the wear increases, the frequency spectrum will change. Looseness severity can range from incipient to severe. Severe looseness can result in rotor instability.
Looseness (structural)	Typically, you will see a strong 1x vibrational component in the spectra, with the highest amplitude being oriented in the direction of lowest stiffness.	An Operating Deflection Shape (ODS) or a Motion Amplification analysis may be helpful in visualizing the vibrational motion of a machine or structure as it's running. Understanding the relative amplitudes and phase relationships between points in the structure helps you understand when happening.

Closing Thoughts

To achieve world-class reliability, repair organizations will need:

1. Maintenance standards that include a comprehensive set of inspection, repair, alignment, and balancing best practices
2. Measuring instruments required to perform the proper component inspections
3. The tools and equipment required to properly perform the repairs
4. The skills required for inspections and repairs, which may require training
5. Software, cameras, calipers, micrometers, etc., to document repair findings
6. Vibration and temperature measuring equipment to gather post repair field data and ensure all indications are safe.

This combination of tools, equipment, knowledge, skills, and a reliability mindset, can transform an average maintenance organization into one that's best in class. However, there may be obstacles to overcome on the journey to world-class reliability.

Sometimes, gray areas are created between reducing downtime and following maintenance best practices, which can force us to make a judgment call between the speed and quality of a repair and the machine's long-term reliability. Pressure from someone in operations to get a machine back on-line can impair our judgment and lead us to cut corners. However, management should always support maintenance personnel when they are attempting to perform the best level of repair based on the situation.

Maintenance organizations should make it a point to a) maintain a set of maintenance best practices, b) make them available to everyone, and c) encourage them to be used consistently and without compromise. I have found it typically makes the most sense to follow maintenance best practices and plant procedures and resist the pressure to ignore your standards in the interest of time. There are no shortcuts to reliability excellence.

References

1. Pekarek, James, *Cutting Corners Can Affect Safety and MTBF*, VIBRALIGN BLOG, December 21, 2016.
2. Perez, Robert and Conkey, Andrew, *Is My Machine OK?*, pp. 225-227.

3. Lifetime Reliability White Paper, *Rotating Machinery Rotor Balancing*
4. Lyons, Jim, *Dynamic Balancing: "Causes, Corrections and Consequences,"* Presented at MainTech South '98, 2 December 1998.
5. Sondalini, Mike, *The Importance of Fit, Tolerance & Clearance*, Accendoreliability.com Website.
6. Efficient Plant Editorial Staff, *Clearance and Interference Coupling Fits*, May 12, 2010.

11

Procedures + Precision = Reliability

By Drew Troyer

In 1963, the USS Thresher, the jewel in the crown of the U.S. Navy's nuclear-submarine fleet, sank during a sea trial after its first shipyard overhaul. We lost 129 brave sailors in that accident. The investigation concluded that relying on "skill-of-the-craft" to guide execution of maintenance work—rather than relying on standardized procedures—led to the errors and defects that caused this tragic mishap. This difficult lesson resulted in the creation of the "Sub-Safe" program, which utilizes procedures and checklists to guide work and minimize human error.

Similar learning has taken place in aviation. As a passenger flying with any major airline today, you would, on average, have to take a commercial flight every day for 25,400 years to die in an aviation crash. Every aspect of aircraft design, manufacture, operation, and maintenance is highly

Robert X. Perez (ed.) Maintenance, Reliability and Troubleshooting in Rotating Machinery, (227–230)

controlled with procedures and checklists. Why are the aviation, nuclear submarine and nuclear power industries so procedure intensive? It's simple: reliability is not a luxury in those industries, it's a mandatory requirement because failure is simply not an option. It should not be an option in other sectors, either.

As a reliability engineering consultant working in the manufacturing and process industries, I review numerous preventive and corrective maintenance plans and operational instructions. I can sum them up in a word: "Vague." I've seen some really bad examples of work instructions that include "inspect the machine," "repair the machine," "check electrical," etc. Consider the examples in Figure 11.1. An instruction that says "check oil pressure" is wrong. We need to clearly define the fit, tolerance, quantity, and quality details that are necessary to execute that PM—we need to define the oil pressure that's acceptable. It could be that oil pressure above a particular value is okay; or that below a particular value is okay; or as is the case in my example, that oil pressure needs to fall in between a specified upper and lower limit.

I've heard people argue that "My tradespeople are well trained and experienced. They should know what to do and what's right." Such logic is fundamentally flawed and simply wrong for the following three reasons:

1. ***Paper Is for Remembering.*** Is it truly reasonable to expect people to remember every torque value for every grade of bolt; every grease quantity of what type of grease is required in the drive-end and non-drive-end of every motor in the plant; the required tension on every v-belt in the plant; the

Wrong	Right
Check oil pressure.	Confirm that the oil pressure is between [X_1] and [X_2] psi.
Check temperature of DE bearing.	Confirm that the temperature of the DE bearing is below [X°] F.
Grease electric motor.	Using [Y] grease, apply [X_1] grams to the DE bearing and [X_2] grams to the NDE bearing of the electric motor.
Tension belt.	Tension belt to [X] lbf, verifying with [Y] tensiometer. Recheck after [Z] operating hours and adjust if required.

Figure 11.1 Example of common mistakes in preventive and corrective work orders.

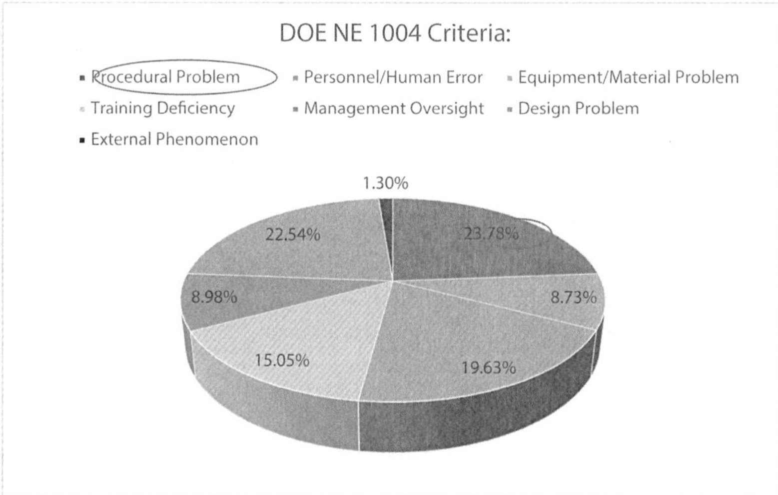

Figure 11.2 My own research suggests that lack of or ineffective procedures is the largest contributor to failures in the plant.

acceptable temperature, pressure and flow ranges or limits for every machine in the plant? While I could go on and on, I think Albert Einstein summed it up well when he said "Paper is to write things down that we need to remember. Our brains are used to think." It's settled science: we need procedure to remember all the details that are required to improve human performance and minimize the human factors of failure, which account for about 80% of all failures in the plant. My own research has concluded that lack of or ineffective procedures are the number-one cause of human failure in the plant (Figure 11.2).

2. ***People + Processes = Success.*** It may well be that your very top tradespeople can work effectively without procedures and checklists. However, there is a reason why even the most experienced airline pilots finger read and completely and faithfully execute their pre-flight inspections and checklists. To that end, even the world's most highly skilled surgeons have acknowledged that they, like all humans, are susceptible to slips and lapses and have turned to procedures and checklists to ensure top performance every time. This is not a question of skill, knowledge, or experience. It is one of discipline and professionalism to counteract distraction

and the possibility of error. The person may receive a call, have a break or get pulled away in the midst of executing a job. Might he or she forget where they left off? Additionally, creating human-performance-management systems that are designed for people whose skills are at the 90th percentile is a near-guarantee of failure. Even managing to the average, or 50th percentile, isn't enough. You must create human-performance-management systems geared to the skills of the 10th percentile to assure success.

3. ***Corporate Amnesia.*** Corporate amnesia is a phenomenon whereby an organization literally forgets how to run its business. How does this occur? Generally, your most senior people are your most experienced personnel; they have most well-developed skill sets for operating and maintaining the plant. If your senior people retire in large numbers, they walk out with the intellectual property that's required to run the plant. Clear procedures geared for the 10th-percentile people convert that tacitly held "tribal" knowledge and know-how into formalized intellectual property. Once formalized, you're in a position to drive consistent quality, manage organizational change and best practice, and enable management of change (MOC) and continuous improvement, continuity, and sustainability.

I'm a big fan of creativity and creative people for analyzing and solving complex problems and innovating new ideas and opportunities. These activities represent the artistic side of plant reliability engineering and management. For day-to-day operation and maintenance of the plant, however, we don't want to rely on the artist's approach. We want to adopt the jet pilot's approach, which is driven by checklists and procedures. I believe that just 10% of the organizational and process rigor employed in the aviation industry would yield 90% in improving the reliability, performance, safety, and environmental objectives we have in a typical plant or factory from managing the human factors of failure.

12

The Top 10 Behaviors of Precision-Maintenance Technicians

By Drew Troyer

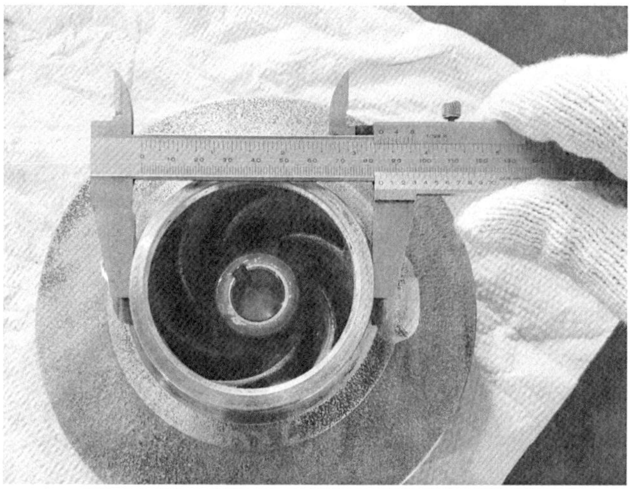

Figure 12.1 Every critical component dimension should be measured and double checked to ensure they fall within the range prescribed by the repair standard.

Executing equipment maintenance is as much about behaviors as it is about skills. Think of a skill as knowledge about HOW to do something, while a behavior is related more to understanding WHY—basically having an appreciation for the motivation.

Behaviors are very much about mind-set and culture. In the plant, we hope to create a culture of defect avoidance through proactive and precision maintenance. This requires a change in behaviors. Here's my top-10 list. Instilling and nurturing these behaviors in precision-maintenance technicians will lead to cultures of proactive and precision equipment care. Note that the following attributes aren't listed in any order of importance.

Robert X. Perez (ed.) Maintenance, Reliability and Troubleshooting in Rotating Machinery, (231–234) © 2022 Scrivener Publishing LLC

1. ***Measure twice, cut once.*** A large percentage of equipment failures are early-life or "infant mortality" failures. These events occur largely due to a lack of precision in the rebuild and/or installation process (see Figure 12.1). The precision-maintenance technician takes care to assure that all controllable quantitative and qualitative aspects of a job are executed correctly by checking and double-checking tolerances and compliance to those tolerances.
2. ***Use the right tool for the right job.*** The precision-minded technician understands that excellent work cannot be performed without the right tools for the job. He or she clearly explains to the supervisor or supporting engineer what tools are required to do the job right and makes sure the job plan specifies those tools and that they are available at the time the job commences.
3. ***Why-why mindset.*** Machines fail for a reason. The precision-maintenance technician understands the physics of failure and the forcing functions that compromise reliability. He or she is constantly asking why a machine failed and could it have been prevented. Likewise, he or she knows which operating conditions induce failure and strives to understand why they've fallen out of control. This understanding allows control to be restored and the cause for the deviation to be eliminated. The precision-maintenance technician is always in root-cause-analysis (RCA) mode—every hour of every day.
4. ***Good parts = good maintenance.*** The precision-maintenance technician understands that we must use the right parts and materials that are in the right condition to ensure proactive control of machine reliability. He or she won't compromise by installing the incorrect part or a part that's been compromised during storage (shelf degradation). Likewise, the precision-maintenance technician makes sure other materials, such as lubricants, are of the right type and that they're clean, dry, and in good condition before being added to the machine.
5. ***A clean machine is a happy machine.*** The precision-maintenance technician is a stickler for cleanliness and order. In the Toyota Production System (TPS), this is often referred to as the 5S system (sort, set-in-order, shine,

standardize, sustain). He or she understands that a clean machine is easier to inspect and, accordingly, find minor defects and conditional abnormalities that can be corrected to restore precision condition control. The precision-maintenance technician also realizes that cleanliness in the plant instills an important sense of pride in the workplace.

6. ***A blood cell-sized lube film separates reliability from failure.*** The precision-maintenance technician understands that a lubricant film of about five micrometers (or about the diameter of a red blood cell) is the difference between machine reliability and failure. He or she is diligent in making sure the lubricant will provide a healthy lubricating film thickness that is free of contamination. The precision-maintenance technician is also rigorous in controlling sources of vibration, including misalignment, unbalance, and looseness, that overwhelm the lubricant's film strength and corrects those issues before they can cause harm.
7. ***I know what's in it for me (WIFM).*** The precision-maintenance technician realizes that in equipment asset-dependent company, such as a manufacturing or process plant, the health of the machinery assets determines the health of the company. He or she understands that if the assets perform well, his or her prospects for continued employment and professional growth are strong, which is good for the entire community. The precision-maintenance technician clearly understands that a reliable plant is a safer and more environmentally sustainable plant.
8. ***Big problems are the by-product of little problems that we didn't address when we could have and should have.*** The precision-maintenance technician understands that ignoring a little problem today results in a big one tomorrow. He or she has learned that machines don't miraculously fix themselves. Thus, the precision-maintenance technician eagerly searches for opportunities to make minor adjustments and corrections so as to proactively prevent issues from turning into big problems later.
9. ***The devil is in the details.*** The precision-maintenance technician demands that preventive and corrective maintenance work instructions and job cards include fit, tolerance, quantity, and quality details necessary to complete the task correctly and with precision. He or she doesn't depend upon

memory or guesswork to execute the job, knowing that lack of (or ineffective) procedures is the leading root cause of machine failure.

10. ***Continuous improvement drives excellence.*** Toyota calls it "kaizen." In English, we refer to it as continuous improvement. The precision-maintenance technician realizes that small, incremental improvements add up and have a major impact on the reliability of equipment assets in a plant. He or she strives to improve parts quality, work instructions, work execution, and work quality assurance every day with the knowledge that "we'll do better" tomorrow and the day after, which, over time, equates to huge gains.

13

Optimizing Machinery Life Cycle Costs through Precision and Proactive Maintenance

By Drew Troyer

Precision Maintenance 101

Smooth, quiet, cool-running machines sum up our goal for precision maintenance. Achieving that goal confirms that we are effectively controlling unproductive uses of energy, which result in friction and manifest as vibration, acoustic emissions, and the generation of heat. Doing so enables us

Robert X. Perez (ed.) Maintenance, Reliability and Troubleshooting in Rotating Machinery, (235–252)

to proactively extend the life of our equipment. Consider these rules of thumb for calculated life-extension potential in mechanical systems:

- Reducing mechanical vibration (velocity in in/sec or mm/s) by 20% doubles the life of mechanical equipment.
- Reducing water contamination in lubricating oil by half extends machine life by 30-35%.
- Cutting particle contamination in lubricating oil by half yields another 20-25% extension in machine life.
- The rate at which chemicals degrade, *g.*, motor insulation, elastomers, etc., is halved for every 10 C-deg. reduction in temperature.
- A 3% phase-to-phase voltage imbalance effectively de-rates a 100-kW motor by 10%, reducing it to a 90-kW-rated motor and increases operating temperature by 18%.

These are just a few of the important relationships between precision maintenance and equipment life. It's simple physics: Control the sources of vibration, heat, abrasion, and material fatigue and deterioration and you'll extend the life of your equipment.

This is not new information, which causes me some concern. Why have so few organizations truly committed to managing the root causes of failure, which are really very easy and inexpensive to manage?

Here is a trial that you might want to consider at your facility to evaluate the benefits of Precision-Maintenance:

1. Select a representative sampling of machines in your plant to serve as the experimental group.
2. Thoroughly baseline the following measurements under defined operating conditions:

    a. vibration readings from multiple locations and components
    b. operating temperature from multiple locations and components
    c. ultrasonic acoustic emission readings from multiple locations and components
    d. lubricant health from oil and/or grease analysis
    e. lubricant particle-contamination levels
    f. lubricant water-contamination levels
    g. energy consumption (amps and volts).

3. Referring to the list of approximately 70 optimum reference states listed in Chapter 14, implement precision maintenance, with a special focus on the following:

    a. Ensure that rotating elements are precisely balanced, ideally to an ISO 1940 G-1.0 level or better.
    b. Precisely align shafts and pulleys, taking into consideration thermal growth and pipe strain.
    c. Ensure that all threaded fasteners are the correct grade, the correct size, properly lubricated, and tensioned correctly utilizing the correct tightening sequence.
    d. Ensure that lubricants are properly selected and applied in the right volume.
    e. Make sure that oils are cleaned before being added to the machine.
    f. For greased components, use grease from tubes because it's typically the cleanest.
    g. Upgrade breathers on tanks to prevent dirt and water ingestion. Desiccant-type breathers may be required.
    h. Install filters to decontaminate the oil. These may be full-time kidney-loop filters, or you may employ quick connects on the machine and a portable filtration device for periodic decontamination. Water-removal technology may be required in certain applications.

4. After stabilizing the experimental group of machines to accurately reflect the improvements made with precision maintenance, retest as specified in item 2 above, and compare new readings to old readings.
5. Apply the following three life-extension equations:

Life-Extension Equations

Vibration Life Extension (based on overall vibration in mm/s):

$$Vibe\ LEF = \left(\frac{New\ Vibe\ Level}{Current\ Vibe\ Level}\right)^{-3.0}$$

Particle Life Extension (based on particles > 10 mm/ml of fluid):

$$Particle\ LEF = \left(\frac{New\ Particle\ Count}{Current\ Particle\ Count} \right)^{-0.35}$$

Water Life Extension (based on water in ppm, which equals percent x 10,000)

$$Water\ LEF = \left(\frac{New\ Water\ (ppm)}{Current\ Water\ (ppm)} \right)^{-0.4}$$

Worked Example

If you've been working the numbers in your head, you can see that a very small reduction in vibration produces a very big impact in life extension as compared to a relatively similar magnitude reduction in water contamination. For example, reducing vibration by half results in an eight times life extension. Conversely, reducing particle contamination by half only results in a 27% life extension, and reducing water contamination by half yields a 33% life extension. These relative relationships can be deceiving. It's rare to reduce vibration in half, and once you decide to clean your oil, you're typically going to reduce particle contamination levels by more than 90%. Consider the life extension calculation found in Table 13.1.

Table 13.1 Life extension example.

Control area	Current value	New value	LEF
Vibration (mm/s)	5	4	1.95
Particle Count (number>10 micrometers/ml)	1,000	30	3.41
Water Content (ppm)	1,000	200	1.9
Total Life Extension			**7.27**
Current MTBF (months)			12
Percent Failures Due to Wear			50%
New MTBF (months)			**44**

If your baseline readings reveal an overall vibration of 5 mm/s, a particle count of 1,000 particles >10μm/ml (approximately ISO 21/19/16) and a water-contamination level of 1,000 ppm (0.1%), and your precision-maintenance experiment drops overall vibration to 4 mm/s, particle count to 30 particles >10μm/ml (approximately ISO 16/14/11), and water content to 200 ppm, your effective life-extension opportunity is about seven times.

Of course, those are estimates for general equipment, as each machine responds a bit differently to precision-maintenance interventions. If, however, you assume that your current MTBF is 12 months and 50% of your events are attributable to wear (this number will also vary), your new effective MTBF is increased to 44 months. This is the power of proactive and precision maintenance.

Life Cycle Costs

All manufacturing processes employ materials, energy, and information to convert raw materials into finished goods that people demand and for which they're willing to pay. The use of materials and energy clearly has environmental consequences. While precision maintenance probably can't substantively impact the raw materials employed in the direct production of whatever the plant makes, it can substantively affect the amount of energy consumed in the direct production process and the energy and materials required to purchase, transport and install machines used in the production process or the parts required to rebuild existing systems.

Increasingly, organizations are expanding the corporate mission beyond profits/shareholder value, safety, and customer satisfaction to include environmental sustainability. Let's explore how precision maintenance can support your goals to reduce your company's environmental footprint and better serve your environmental sustainability mission objectives.

Considering Energy Consumption

Friction is the difference between machines that run loud, rough, and hot and those that run quiet, smooth, and cool. Precision maintenance takes dead aim at the sources of friction, i.e., mechanical looseness and poor electrical conductivity, unbalance (mechanical or electrical), misalignment, and poor lubrication. Mechanical friction and electrical resistance (I^2R losses) are parasitic uses of energy that manifest in the form of vibration, noise (acoustic emissions) and heat.

Noted tribologist Peter H. Jost suggested that reducing friction through proper lubrication practices alone can reduce energy consumption by a whopping 7.5%. Some experts suggest that precision shaft alignment alone can reduce energy consumption by as much as 10%. Comparing the loudest, roughest, and hottest machines to those that are the quietest, smoothest-running, and coolest, I estimate that we conservatively have a 5% to 15% energy- consumption-savings opportunity for typical plant-equipment types. This has both economic and ecological implications.

Let's first look at the economic implications. Suppose that a 100-kW electric motor runs 90% loaded over a 7,000-hr. production year. That motor utilizes 630,000 kWh of electricity per year. Assuming electricity costs $0.07 per kWh, the annual cost to energize the motor is $44,100. Suppose that, through precision maintenance, we can reasonably reduce friction related energy losses by 5%. Our savings would be $2,205 per year for the *one* electric motor. Now, let's say that the plant has 25,000 kW of power deployed at 90% load for a 7,000-hr. production year, and we can reduce frictional energy losses by 5% plant-wide. The annual savings would be $551,250. Consider these simple examples, then plug in your own numbers.

The dollar savings associated with reducing frictional losses with precision maintenance go straight to the bottom line. There's more to consider. The generation of power releases carbon dioxide and other greenhouse gases (GHGs), which contribute to climate change. The U.S. Environmental Protection Agency (EPA) estimates that there is 0.707 kg of carbon dioxide (or its equivalent in other GHGs) released for every kWh of electricity. So, using our example of a 100-kW motor operating at 90% load for a 7,000-hr. year emits 445.41 metric tons of carbon dioxide, or its equivalent, per year. Reducing frictional losses by 5% reduces our carbon footprint by 22.27 tons per year. Expanding the analysis to our example of a plant deploying 25,00-kw of power at 90% load over a 7,000-hr. production year, a 5% reduction in frictional losses results in a 5,567.63 metric ton reduction in GHG emissions.

So, what does this mean to you? First, these are numbers that you can report to the public, particularly if you've taken my challenge to measure energy consumption before and after deploying precision maintenance. It's good public relations (PR) and good marketing as a greater percentage of the population becomes concerned about anthropogenic climate change. However, there could be a financial impact as well. Opinions vary about the social costs of carbon (SCCs), i.e., the social costs associated with each ton GHGs emitted and presented and carbon-dioxide equivalent units. The most commonly cited figure is $45 per ton.

Table 13.2 Economic and environmental impact analysis associated with reduction of frictional energy losses through proactive maintenance.

	100 kW motor	**25,000 kW plant**
Load Factor	90%	90%
Operating Hours	7,000	7,000
kWh Per Year	630,000	157,500,000
Electricity cost (per kWh)	$0.07	$0.07
Annual Electricity Cost	$44,100.00	$11,025,000.00
Annual GHG Emissions (metric tons of carbon dioxide)	445	111,353
Reduced Frictional Loses from Precision Maintenance	5%	5%
Energy Savings:		
kWh Per Year	31,500	7,875,000
Dollars	$2,205.00	$555,250.00
Greenhouse Gas Reductions:		
Metric Tons of Carbon Dioxide	22	5,568
Social Cost of Carbon ($45 per metric ton)	$1,002.17	$250,543.13
Total Economic Benefit	$3,207.17	$801,793.13

The idea of a carbon tax continues to gain traction globally and has already been implemented in some areas (e.g., California). If the concept of a carbon tax catches on, the previously discussed public-relations benefits become real dollars. For example, assuming $45 per metric ton, a 5% reduction in frictional losses through precision maintenance is worth a little over $1,000 per year for our 100-kW-electric-motor example, and more than $250,000 per year for our 25,000-kW plant-wide example. Keep in mind that those types of SCC savings are in addition to the actual economic savings previously discussed. All figures for the economic and environmental impact analysis are summarized in the Table 13.2.

Life Cycle Inventory Analysis

When machines run quietly, smoothly, and cool, they not only consume less energy, but they also last longer because less friction means less wear. In the example from the first installment, we estimated that by focusing on Fasteners, Lubrication, Alignment and Balance (FLAB), the foundation of precision maintenance, we could increase the MTBF by a factor of 3.66 (from 12 to 44 months in our example). This type of result has significant environmental implications.

In the field of environmental sustainability, we employ a process called Life Cycle Inventory Analysis (LCIA). The purpose of LCIA is to calculate the overall environmental impact of the acquisition of materials, production, use, recycling/reuse and waste treatment, and required energy and transport associated with the creation and use of goods and services that consumers demand (Figure 13.1). The LCIA process is defined in ISO Standards 14040 and 14044. These ISO standards are essential elements of ASTM Standards for Sustainable Manufacturing that have been created by the ASTM E60.13 committee (of which I'm a voting member). The current ASTM Standards for sustainable manufacturing include E29816-15, E3012-16, E2987/E2987M-16, and E3096-18. More related standards are under development.

While it's beyond the scope of this chapter to teach the LCIA process, we can simply conclude that if we calculate the environmental LCIA impact associated with the acquisition or rebuild of any component or system AND we increase the time between replacements from 12 to 44 months, we will have fundamentally decreased our environmental impact by 73% (excluding direct energy use associated with the machine's use, which we addressed previously). Adding this to the previously discussed energy savings, precision maintenance can have an enormous positive environmental impact.

Justifying Precision Maintenance

Estimating the Benefits

Before diving into our precision-maintenance-justification example, let's discuss some fundamentals of cost-benefit analysis for investments in asset reliability improvement. Referring to Table 13.3, there are several principles to observe:

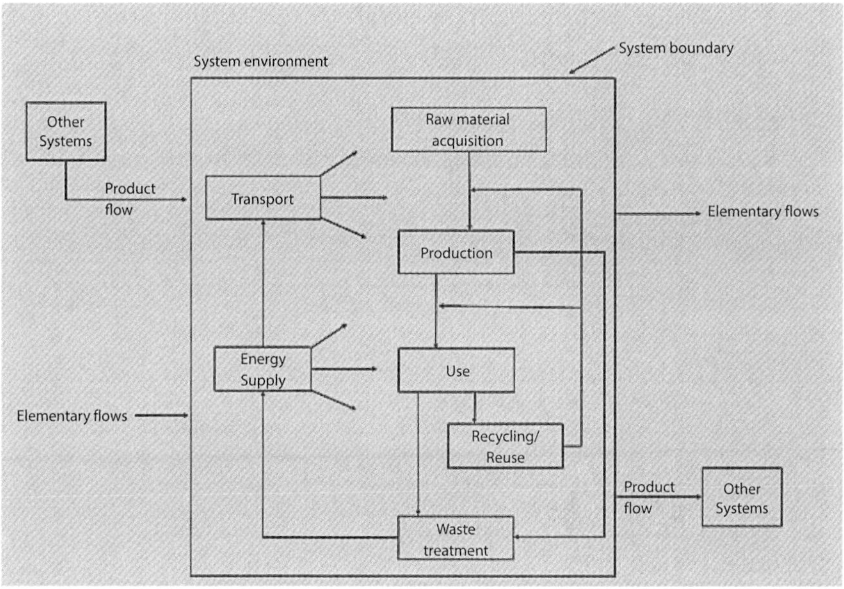

Figure 13.1 Life cycle inventory analysis (LCIA) flow diagram, per ISO 14040.

1. The true cost of a failure can't be deterministically defined until AFTER the failure event, when it's too late.
2. Efforts to detect failures, (e.g., predictive monitoring), prevent collateral damage, and plan and schedule corrective interventions will reduce the severity of each failure event.
3. Efforts to design for reliability, maintainability, and operability, operate the equipment within acceptable levels, and proactively maintain the equipment to reduce vibration, noise, and heat (e.g., precision maintenance) will reduce the number of failures per year.

Table 13.3 Notes:

1. Planning tools affect the distribution by severity levels:

   a. Protective monitoring
   b. Predictive monitoring
   c. Planning and scheduling

Table 13.3 Investments to improve the reliability of plant assets focus on reducing the severity of each failure event and/or reducing the number events that occur per year. Often, you can achieve both. *(Ref: Plant Reliability in Dollars & $ense Workshop – Drew D. Troyer, CRE).*

The true cost of a failure can't be estimated until after the fact.

For planning purpose, a probabilities approach must be utilized.

	Failure severity			
	High	**Moderate**	**Low**	
Cost Category				
Parts	$ 2,000	$ 1,000	$ 1,000	
Labor	$ 2,000	$ 1,000	$ 1,000	
Downtime	$ 10,000	$ 2,000	$ -	
Risk-based	$ 1,000	$ 500	$ 200	
Total	$ 15,000	$ 4,500	$ 2,200	
Likelihood	0.1	0.2	0.7	Total
Weighted Cost	$ 1,500	$ 900	$ 1,540	$ 3,940
Failures/Year				2
Annual Failure Cost				$ 7,880

2. Proactive measures influence the failure rate:

    a. Design for reliability, maintainability, and operability
    b. Condition control
    c. Precision control
    d. Precision maintenance

Unless you have a crystal ball, forget about getting perfect data to predict the economic benefits of an investment in asset reliability. While we employ as much deterministic and empirically derived data as is available, at some point we need to move forward using the best information at hand. We must simply get comfortable with the fact that we don't always have perfect data. Our predictive monitoring tools and maintenance-work management methods for planning, scheduling, and coordinating work reduce the severity of each failure event, but they don't necessarily reduce the number of failures we see over a given amount of time.

Consider Table 13.4. In this example, we estimate that the failure cost for a machine has three severity levels: high, moderate, and low. The high-severity event carries the greatest cost, which is primarily attributable to downtime. The moderate scenario has similar repair costs, but a much lower impact on productivity. The low-severity scenario has lower repair costs and no impact on downtime.

In our example, we estimate that 10% of the events are severe and 30% are moderate, which leaves minor events making up 60% of the failure scenarios. Calculating the weighted average cost per event, we estimate the current annual unreliability cost to be $4,270.

Let's assume that we implement vibration analysis, oil analysis, thermography, ultrasonic analysis, sensory inspections, or other methods to provide an early indication that the machine has a fault and a failure is imminent; and that we employ planning and scheduling tools to ensure the necessary parts and tools are on hand, maintenance instructions are created, and machine access is negotiated with production. We estimate that these interventions will rearrange the severity distribution so that only 3% of the events are high severity, 5% are moderate severity and 92% are low severity, reducing the weighted average cost per event to $2,627—a savings of $1,643 per event—which is a reduction of 38%.

Earlier in this chapter, I suggested a scenario of proactive intervention for reducing vibration, noise, and heat generation by focusing on the mechanical and electrical aspects of fasteners, lubrication, alignment and balance (FLAB). In that example scenario, we increased the mean time between failure (MTBF) from 12 to 44 months (see Table 13.1). This proactive intervention reduces the failure rate from 1.0 to 0.27 per year. Assuming that we implement both the predictive and proactive (precision) interventions, in our example, the annual failure cost is reduced from $4,270 to $709—a significant improvement.

Now for the Cost-Benefit Analysis

It's not enough to estimate the benefits of an investment in asset reliability improvement. It's got to make sense on the bottom line when we compare the benefits to the cost. An important concept to cost-benefit analysis is the time value of money. In essence, a dollar is worth more today than at any point in the future, and the discount of a future cash flow depends upon the cost of capital. The present value (PV) calculation

Table 13.4 Summary benefits analysis for a reliability-improvement initiative.

	Failure severity			
Costs	*High*	*Moderate*	*Low*	
Parts	$ 2,000	$ 1,000	$ 1,000	
Labor	$ 2,000	$ 1,000	$ 1,000	
Downtime (less COGS)	$ 10,000	$ 2,000	$ -	
Other	$ 500	$ 200	$ 100	
Total	$ 14,500	$ 4,200	$ 2,100	
Percentage Distribution				
Current Percentage	10%	30%	60%	
Proposed Parentage	3%	5%	92%	
Weighted Avg. Failure Cost Per Event				
Current Percentage				$ 4,270
Proposed Parentage				$ 2,627
Failures Per Year				
Current Situation				1
Proposed Situation with Precision				0.27
Annualized Failure Cost				
Current Situation				$ 4,270
Proposed Situation with Precision				$ 790

example below illustrates how $1,000 received five years in the future, at a cost of capital of 10% per year, has a present value of $621 in today's dollars.

$$PV = Cash\ Flow / (1+k)^n$$

Where:
PV = Present Value
Cash Flow = The amount received (or avoided) or the amount spent at a point in time
k = Discount rate (usually inflation adjusted)
n = number of periods or years in the future
Example: $1000 received 5 years in the future at a capital cost of 10%

$$PV = \$1000 / (1+0.10)^5 = \$621$$

A dollar is worth more today than at any time in the future.

Source: *Plant Reliability in Dollars & $ense Workshop* - Drew D. Troyer, CRE.

In our example, we have estimated that we can reduce our annual unreliability cost from $4,270 to $709 for a savings of $3,561—that's the benefit. However, we must consider a number of costs. Such costs could include:

1. equipment modifications to reduce vibration, lubricant contamination and to improve lubrication effectiveness
2. tools required for machine condition monitoring and precision maintenance (apportioned)
3. investments in software and people to improve work management, parts quality management, etc. (apportioned)
4. training, coaching, and mentoring support to produce a culture change (apportioned)
5. reliability engineering (apportioned)
6. management and supervision (apportioned).

Let's make the following assumptions in our example:

1. We have 20 machines of the type included in our analysis.

2. The upfront cost to modify each piece of equipment for precision maintenance is $2,500.
3. The ongoing cost to support the modifications is $250 per unit per year.
4. The apportioned upfront cost for tools, training, systems, engineering, and management support is $500 per unit.
5. The apportioned ongoing cost for tools, training, systems, engineering, and management support is $100 per unit per year.
6. First-year benefits will be discounted by 50%. Second-year benefits will be discounted by 25%. We sometimes discount the first two years' benefits to reflect the time required for full implementation of the changes.
7. We're conducting a five-year cost-benefit analysis.
8. Management is charging us a 12% cost of capital to underwrite the investment.
9. Management has requested that we limit our economic analysis to the hard benefits, but suggested that we include the soft benefits, such as reducing GHG emissions and the associated social cost of carbon in our report.
10. Management has requested that we analyze the hard benefits utilizing the following:

    - Net present value (NPV) —the total value of the project in today's dollars.
    - Internal rate of return (IRR) —the effective interest rate of the investment's return.
    - Discounted payback period—the time required to recover the invested dollars using discounted future cash flows.

Now that we have our assumptions and our marching orders from management, we're in a position to tabulate our cost-benefits analysis. The findings of our cash-flow analysis are summarized in Table 13.5 below.

Our proposed project provides a healthy five-year NPV of over $600,000. This number represents the increased profit to the firm over the first five years of the investment's life discounted to reflect the value of money today. That's not to say that our precision-maintenance intervention won't continue to yield benefit after five years—there's every reason to believe that it will. However, we're limiting our planning horizon to five years because the future can be very uncertain. The IRR is 220%, which means that we would have needed to find a bank that could pay us an annual interest rate of

Table 13.5 A summary of our cost-benefit analysis.

	Year				
	1	2	3	4	5
Benefits (for 20 units):					
Reduced maintenance and reliability costs	$0.00	$35,610.00	$53,415.00	$71,220.00	$71,220.00
Reduced energy costs	$0.00	$22,050.00	$33,075.00	$44,100.00	$44,100.00
Total annual benefits	$0.00	$57,660.00	$86,490.00	$115,320.00	$115,320.00
Upfront costs (20 units):					
Engineering Modifications	-$50,000.00				
Engineering, training, and systems	-$10,000.00				
Upfront costs (for 20 units):					
Equipment modifications		($5,000)	($5,000)	($5,000)	($5,000)
Engineering, training, and systems		($1,000)	($1,000)	($1,000)	($1,000)

(Continued)

Table 13.5 A summary of our cost-benefit analysis. (*Continued*)

	Year				
	1	**2**	**3**	**4**	**5**
Net cash flow	($60,000)	$109,320	$166,980	$224,640	$224,640
Discount factor (k=12%)	1.00	0.89	0.8	0.71	0.57
Discounted cash flow	**($60,000.00)**	**$97,294.80**	**$133,584.00**	**$159,494.40**	**$128,044.80**

Summary Investment Analysis	
5-year net present value (NPV)	$600,846
Internal rate of return	220%
Discounted payback period	0.61 years

220% to match the quality of this investment. Finally, this discounted payback period is 0.61 years, which means that we'll recover all the invested money in 7.3 months—a pretty good investment indeed.

In the analysis report to management, we'd certainly highlight the fact that we believe that our precision-maintenance initiative is expected to reduce our carbon footprint by 440 metric tons of CO2 per year—a reduction in the social cost of carbon of more than $20,000 per year. However, we frequently make note of these types of improvements in the report as positive externalities, while excluding them from the quantitative cost-benefit analysis.

I would like to conclude this chapter by encouraging readers to take my precision maintenance challenge and test opportunities for increasing equipment life, reducing maintenance and operating costs (including energy costs), and minimizing your organization's environmental footprint through precision maintenance. I recommend that you use Chapter 14 titled "Optimum Reference States for Precision Maintenance" as a basis for optimizing your precision maintenance programs, with regard to fasteners, lubrication, alignment, and balance (FLAB).

14

Optimum Reference States for Precision Maintenance

By Drew Troyer

Figure 14.1 Precision maintenance is one of the hallmarks of world-class machinery reliability.

Of all the various strategies associated with the maintenance and reliability of production equipment in the manufacturing and process industries, none is more salient than precision maintenance when it comes to increasing production availability and reducing life-cycle ownership costs of assets (see Figure 14.1).

Precision maintenance focuses upon managing the root causes of failure. When properly implemented, it results in quieter-, smoother-, and cooler-running equipment. In addition to improving equipment uptime, reducing maintenance and life-cycle ownership costs, and extending the

Robert X. Perez (ed.) Maintenance, Reliability and Troubleshooting in Rotating Machinery, (253–262)

life of assets, precision maintenance decreases energy costs and associated greenhouse gases (GHGs) by targeting and eliminating sources of friction.

I've organized my approach to precision maintenance into what I call "FLAB Management." FLAB is the acronym for and electrical and mechanical **F**asteners, **L**ubrication, **A**lignment, and **B**alance—in other words, the foundation of electromechanical reliability. As listed here, I've created about 70 Optimum Reference State (ORS) conditions associated with the four elements of FLAB and the overarching category of FLAB management.

Each ORS statement is written as a positive statement to enable you to evaluate your organization's performance. These statements certainly don't reflect a complete or comprehensive definition of the entire universe of precision maintenance. They do, however, represent a credible cross-section. Use them to gauge your operation's relative strengths and weaknesses in precision maintenance.

Fasteners

- We include torque values, fastener lubrication requirements, and required fastening sequence in all corrective- and preventive-maintenance-work instructions.
- We employ properly selected and specified (e.g., grade, material, etc.) and sized fasteners and washers for fastener applications, and the fastener type, size, and material are clearly defined in maintenance-work instructions.
- Fasteners are audited at least twice a year, and a routine audit of fasteners will reveal no loose, missing, improperly sized, or damaged fasteners in the plant.
- Torque wrenches are employed for tightening fasteners in the plant, and when a specialized torque wrench is required, it is specified in the maintenance-work instructions.
- Foundation-mounting bolts are properly selected, sized and grouted, and machines are correctly shimmed, when installed, to eliminate soft foot.
- Belts on belt-driven equipment are properly tensioned with a spring or vibration tensiometer upon installation, re-tensioned shortly after start-up to adjust for relaxation or belt seating and required tension values, and tensioning instructions are explicitly included in corrective- and preventive-maintenance-work instructions.
- Belts are inspected visually (or with a stroboscope), at least once a month, and checked for tension at least once a year.

- Ultrasonic analysis is routinely employed for finding, tagging, and correcting air and other pressurized gas leaks. Leaks are fixed as found and recorded where possible.
- Machines are routinely inspected for liquid leaks. Leaks are found, tagged, and corrected. Where required, fluorescent dye is employed to identify the location of leaks. Leaks are fixed and found and recorded, where possible.
- All mechanical crafts people are properly trained and qualified on the theory and practice of applicable forms of fasteners, belt installation and tensioning, and leak detection and management. Where appropriate, specialized monitoring is contracted out to qualified experts.
- Mechanical looseness is routinely monitored utilizing vibration analysis, where appropriate.
- Looseness or soft foot issues revealed by vibration analysis or other inspection techniques are attended to with a high priority, before damage to bearings, gears, and other components can occur.
- Electrical connections, including circuit breakers, fuses, contactors, overloads, disconnects, lug connections, etc., are routinely tested with thermography, motor analysis or other inspection techniques.
- All electrical craft persons are properly trained and qualified on electrical-fastener theory and practice, to include monitoring and inspection techniques. Where appropriate, specialized monitoring is contracted to qualified experts.
- Electrical fastener and circuit problems revealed by thermography, motor-circuit analysis, or other techniques, are corrected promptly before they can lead to equipment damage and/or functional failure.

Lubrication

- The required viscosity grade and viscosity index have been analyzed for each application. The analysis considers the operating temperature and range. Viscosity and base-oil type requirements are written into material-specification standards for greases and oils.

- The required additive system has been evaluated relative to the performance requirements of the application. Additive requirements, along with associated performance-property requirements, are written into material-specification standards for greases and oils.
- The appropriate grease thickener has been selected for grease-lubricated applications, and measures have been taken to minimize cross-contamination of thickeners (e.g., instructions to motor-rebuild shops specify the exact grease to use for initial fill).
- The required re-lubrication interval has been technically evaluated for each application to consider component type and size (e.g., bearings), operating speed, vibration, contamination in the area, shaft orientation, operating temperature, leakage rate, etc. The analysis produces ideal re-grease and oil-change intervals, where condition monitoring is not employed to determine the interval.
- The required re-lubrication volume has been calculated for greased bearings considering the component size and type (e.g., bearing), seal configuration, etc., and precision methods are employed to ensure that the correct volume is applied.
- Oil-lubricated machines are equipped with non-intrusive means to check oil level, and these levels are monitored and adjusted, as required, at least once per week (more often where leakage is common).
- Particle-contamination-control limits have been set for all oil-lubricated machines, and proper measures have been taken to exclude and remove particles as required to achieve those targets. Oil analysis is employed as the feedback mechanisms to ensure that targets are achieved.
- Water-contamination-control limits have been set for all oil-lubricated machines, and proper measures have been taken to exclude and remove water contamination, as required, to achieve those targets. Oil analysis is employed as the feedback mechanisms to ensure that targets are achieved.
- Lubricant-storage containers, transfer devices, lubrication tools, and machines are all fitted with intuitive tagging labels (e.g., shapes and colors) to identify lubricants and to avoid cross contamination.

- Lubrication PMs are written clearly to specify proper methods for carrying out various lubrication tasks, and fits, tolerances, quantity, and quality details are included in the associated maintenance-work instructions.
- Lubricant samples are drawn at the proper interval, from the proper location, and using appropriate methods to assure representative oil-analysis data, and the samples are tested to an appropriate test state, for which action limits and thresholds have been set.
- All crafts persons are properly trained and qualified on the theory and practice of lubrication and oil analysis. Where appropriate, specialized testing and analysis is contracted out to qualified experts.
- Transformer oils are correctly managed, maintained, and tested to ensure the necessary dielectric performance and to avoid the accumulation of potentially flammable, dissolved gases, such as acetylene.
- All crafts persons are properly trained and qualified on the theory and practice of lubrication and oil analysis. Where appropriate, specialized testing and analysis is contracted out to qualified experts.
- Lubrication, contamination or wear issues revealed by oil analysis, vibration analysis, or other inspection techniques are attended to with a high priority, before damage to bearings, gears, and other components can occur.

Alignment

- Shaft- and sheave-alignment-maintenance-work instructions include allowable misalignment, so there is no guesswork for techs at the job site. Tolerances are calculated based on speed and consider shaft length when calculating angular misalignment limits.
- Thermal growth is considered when setting up alignment limits and maintenance-work instructions.
- Laser-alignment tools are employed for aligning shafts and sheaves.
- Flexible couplings are NOT used as an excuse to employ lazy and imprecise shaft-alignment practices.
- Pipework is installed properly to minimize pipe strain. Thermal growth is considered when laying in pipework.

Also, piping is properly mounted and secured to reduce movement-induced stress on joints and flanges.

- All mechanical-craft persons are properly trained and qualified on alignment theory and practice for shaft- and sheave-alignment jobs. Where appropriate, specialized work, monitoring and testing is contracted out to qualified experts.
- Mechanical misalignment is routinely monitored utilizing vibration analysis where appropriate.
- Mechanical misalignment issues revealed by vibration analysis or other inspection techniques are attended to with a high priority, before damage to bearings, gears, and other components can occur.
- Electrical-total-harmonic distortion (misalignment with sinusoidal power wave) is maintained to below 3% for electric-motor applications.
- The presence of stray voltage is routinely monitored in equipment, where appropriate (e.g., motors, generators, panels, etc.). Stray voltage is the accumulation of electrostatic electrical potential. When the accumulation reaches a critical level, the potential is electrokinetically discharged, causing electrical-discharge erosion (fluting) and the potential for injury.
- All electrical-craft persons are properly trained and qualified on theory and practice related to managing total-harmonic distortion in electric motors. Where appropriate, specialized monitoring and testing is contracted out to qualified experts.
- Electrical-harmonic distortion and stray voltage are routinely monitored utilizing motor analysis and other technologies, where appropriate.
- Electrical-harmonic distortion and stray voltage issues revealed by motor analysis or other inspection techniques are attended to with a high priority before damage to motors and motor-control centers (MCCs) can occur.

Balance

- Where appropriate, pumps, blowers, fans, etc., are shop-balanced to appropriate standards prior to being put into service.

- Required balancing precision is included in contracts whenever equipment is rebuilt offsite (e.g., electric motor rebuild shops). Where required, at-speed (high-speed) balance is specified.
- Where appropriate, machines that can be field-balanced are, when required. Fans and other air-handling units are a common example of such equipment.
- When correcting balance problems in the field, care is taken to minimize the risk that the cure won't create other problems. A common example is the wash-down of fan blades introducing water contamination into bearings.
- All mechanical-craft persons are properly trained and qualified on dynamic-balance theory and practice and belt installation and tensioning theory and practice. Where appropriate, specialized work and testing is contracted out to qualified experts.
- Mechanical imbalance is routinely monitored utilizing vibration analysis, where appropriate.
- Mechanical-imbalance issues revealed by vibration analysis or other inspection techniques are attended to with a high priority, before damage to bearings, gears, and other components can occur.
- Phase-to-phase-electrical-voltage imbalance is monitored and maintained to acceptable levels to assure maximum motor life. Failure to maintain phase-to-phase-voltage balance results in the generation of heat. Voltage imbalance should be managed to below 2%.
- Electrical-current imbalance is monitored and maintained to acceptable levels to assure maximum motor life. Failure to manage current imbalance causes heat generation and can stress electrical circuits. As a rule, current imbalance will be about seven times higher than voltage imbalance. However, current imbalance can be caused by the circuit, even if phase-to-phase voltage is in balance.
- Phase-to-phase electrical-inductive imbalance caused by poor rotor condition is monitored and maintained to acceptable levels to assure maximum motor life. Inductive imbalance is an indication of wind quality. Limits are 7% for form-wound motors and 12% for loose-wound motors. The lower the better. This is an acceptance criterion for new or rewound motors.

- Electrical-resistive imbalance is monitored and maintained to acceptable levels to assure maximum motor life. Resistive imbalance is a proactive indicator and often a precursor to current imbalance.
- All electrical-craft persons are properly trained and qualified on electrical-balance theory and practice. Where appropriate, specialized work, monitoring, and testing is contracted out to qualified experts.
- Electrical imbalance is routinely monitored utilizing motor analysis (current and circuit), where appropriate.
- Electrical-imbalance issues revealed by motor analysis or other condition-monitoring or inspection techniques are attended to with a high priority, before damage to motors, MCCs, and other components can occur.

Flab Management

- As a matter of policy, proactively managing FLAB with precision-maintenance practices is a priority for management and not flippantly disregarded in lieu of fixing broken equipment.
- Organizational roles pertaining to FLAB management and execution have been clearly defined and are visible on a responsible, accountable, consulted, or informed (RACI) chart. Individuals have received adequate FLAB education and training and are properly supported to fulfill their respective roles.
- All preventive- and corrective-work practices are documented to identify best practice and to specify fits, tolerances, quantity, and quality details specific to each machine and application to ensure precision in the maintenance process without depending on "tribal knowledge."
- Inspections and condition monitoring (e.g., vibration analysis, oil analysis, etc.), as opposed to functional failure, drive the work request, planning, and scheduling process. And condition-directed work is given a high priority so work requests don't just sit in a backlog until the unit reaches functional failure.
- The organization has a good balance of leading and lagging indicators that drive behaviors. For example, overall lubrication effectiveness (OLE) and overall vibration effectiveness

(OVE) are leading indicators. Reliability, availability, cost per ton, etc., are lagging indicators.

- Rewards are tied to achieving proactive goals (e.g., OLE and OVE), not just production goals or reacting effectively. Remember, rewards can be extrinsic (such as money), or intrinsic (such as recognition for effort). Historically, we've rewarded reacting to failure with overtime and pats on the back, rather than reward proactive behaviors that drive reliability.
- The potential benefits in terms of avoided maintenance cost, increased production availability and utilization, and improved safety have been systematically analyzed and quantified in economic terms.
- The organization employs world-class work management practices to ensure that proactive FLAB-related PMs and work that is identified and requested is properly planned, scheduled, and executed prior to damage occurring to the equipment.
- Managers and supervisors have been trained and qualified on world-class asset management (e.g., to the five pillars of the Body of Knowledge set forth by the Society for Maintenance & Reliability Professionals [SMRP]).
- The organization has access to a qualified reliability engineer who is an expert in all aspects of FLAB management, data mining and analysis, maintenance-work management, and other aspects of reliability engineering and equipment-asset management.

Conclusion

If your facility is serious about increasing uptime, extending equipment life, and reducing maintenance and operations costs of production assets, then precision maintenance should be a focal point. That's where reliability hits the plant floor. A "Focus on FLAB" is a great way to organize your efforts in targeting the root causes of friction, wear, and premature failures. It's my sincere hope that this group of Optimum Reference States for precision maintenance will serve as a starting point for you to assess your organization's relative strengths and weaknesses and, from there, formulate an action plan to leverage the opportunity to put precision maintenance to work for your operations.

15

Writing Effective Machinery Work Order Requests

By Drew Troyer

When writing work order requests, operators should attempt to describe the field machinery problem they have encountered.

Figure 15.1 Operator inspecting a pump in the field.

It is vital that process operators closely monitor their process machinery in order to identify machinery failures in their earliest state so that costly secondary damage due to catastrophic failure can be minimized (Figure 15.1). To perform this function effectively, they must understand their machines and have a good feel for what is normal performance and what is abnormal performance. As they walk around, they need to constantly look for leaks and be aware of unusual odors or sounds that may be present. Machinery that is deteriorating will manifest its distress by either running hotter than normal, emitting smoke or vapors, vibrating, or developing

Robert X. Perez (ed.) Maintenance, Reliability and Troubleshooting in Rotating Machinery, (263–268)

spots of burnt paint; all these symptoms should be investigated further. If the operator is convinced there is a problem, they may need to consider writing a work order request to initiate a maintenance work order.

Here are a few tips for defining a field problem on a work request:

- Clearly define the symptoms, i.e., "The compressor is leaking, vibrating, losing flow, losing pressure, etc."
- Explain timing, i.e., a) We noticed the pressure dropped suddenly last night, or b) The night shift noticed a loss of flow this morning."
- Provide precise location information, i.e., "The issue seems most prominent at the inboard end of the west charge pump."
- Explain any observed trends, i.e., "It seems to be gradually getting worse over the last few days."
- Name of operator who knows the most about the problem.

Remember that if you only describe the symptom(s) without context, only the symptom(s) will be treated. Here are some examples of how only treating systems could be a mistake:

- If the symptom reported is a hot bearing, the recommendation may be to increase the water flow to provide additional cooling. However, the problem may be related to a failing bearing, oil contamination, or a change in process load. Here is an example of context information: "Due to process requirements, we have been running the compressor close to its maximum speed." This additional information provides additional information to help the mechanics troubleshoot the problem.
- If the symptom reported is bearing vibration, the recommendation may be to add brace under bearing housing to reduce vibration. However, the problem may be related to an internal machinery issue. Here is an example of context information: "Last night there was an upset in the main column. We think some tray parts may have come loose and entered one of the bottoms pumps." This additional information helps the vibration analyst troubleshoot the problem.

Tip: Provide context, timing, background, etc., to allow the machinery professional to focus on the root cause of the problem.

Here are some bad examples of work order requests:

- Replace seal
- Come look at pump
- Pump sounds funny

To improve your work requests, try to provide specific information to the planner. Here are some examples of vague versus specific work request information:

	Vague	Specific
1.	One of the operators	An operator on A shift noticed
2.	It's been going on for a while	The problem's been getting worse over the last two weeks
3.	One of the charge pumps	The west charge pump
4.	The pumps been acting up	We've seen a 20% loss of flow

Useful work order requests tips:

- Assumes the reader knows nothing about the problem. Improved communication saves time and eliminates errors.
- Try to write in complete sentences.
- Don't tell mechanics how to handle the problem. Instead, keep an open mind when providing details about the issue at hand and don't lead the reader to a conclusion or specific work scope. Let the maintenance department personnel troubleshoot the problem independently and allow them to do their jobs. Maintenance personnel provide an unbiased second opinion about the nature of the problem.
- Provide additional details, such as system details, timeline events, and production circumstances maintenance personnel may not be aware of.

Good examples of work requests:

- The north charge pump's outboard seal oil pot level has fallen below the "fill line." The "A Shift" reported that the level dropped overnight. Contact R. Smith for more information.

- The "B Shift" reported that the flow on the south charge pump has fallen off suddenly. They also reported that the driver speed has gradually fallen off over the last few days. Contact S. Jones for more information.
- The "A Shift" reported that the flow from the north recycle pump dropped off suddenly overnight. Upon inspection one of the shift operators, T. Rodgers, noticed the suction pressure is now running 10 psi below the normal pressure.
- The "B Shift" reported that the ID fan seems to be vibrating excessively. One of the operators, L. Blevins, reported the vibration levels seem to jump immediately soon after start-up after routine maintenance.
- "A Shift" has noticed that recently the recycle compressor's gas turbine driver is losing speed every day during the heat of the day. The loss in compressor speed is now affecting production. Contact M. Wells for more information.
- The "B Shift" has reported that the north make-up compressor vibration level runs above the alarm setting at speeds above 8000 rpm. Due to increased charge rates, we need to run the compressor faster. A trend of compressor speed shows we are running faster that we have ever run it before. Contact J. Johns for more information.
- One of the "A Shift" outside operator noticed the south recycle compressor discharge piping is now vibrating excessively. Upon closer inspection, he observed that the piping support structure is severely cracked. Contact N. Fielder for more information.

Be sure to include any steps you may have taken to eliminate the symptoms or to better understand the problem, such as:

- Cleaning strainers
- Checking instrument calibrations
- Switching pumps
- Changing the machine's speed to see the effect
- Taking amperage reading

Any initial troubleshooting steps already taken will save time by providing insight into the problem and possible causes that can be checked off the list of potential issues by the next group of troubleshooters.

As I've explained in this column, the key to a good work request is providing details, details, and more details. Providing details allows mechanics to properly troubleshoot the problem at hand. For example, if the problem appears after a long machine run, then normal internal wear-out is a real possibility; but if the problem appears immediately after a repair, then perhaps a repair error is the source of the problem. Another example of machinery troubleshooting logic is if a problem occurs immediately after a process upset, then perhaps internal damage due to off design conditions is a possibility.

Effective writing is the means to express language and thought in a way that allows the reader to clearly understand what the writer is saying. Therefore, in order to write effectively, operators need to understand both the complex ideas that they would like to convey as well as the necessary grammatical structures needed to convey them effectively. We should all strive to improve our communication written skills through practice and feedback. An occasional in-house class for operators can also help hone their written communications skills.

In summary, I recommend that you always try to write work request in complete sentences with as much detail as required to explain the problem. Let the planner know what the problem is, where it is occurring, when did it or does it occur, and who knows the most about the problem. The more details the work request contains, the less time the planner and mechanics will have to spend investigating the issue and preparing for the work. Poorly written work requests can lead to confusion and can possibly result in higher repair costs due to multiple trips to the field or rework.

Part IV
ANALYZING FAILURES

16

Improving Machinery Reliability by Using Root Cause Failure Analysis Methods

By Robert X. Perez

Figure 16.1 Premature machinery failures result in higher maintenance costs and lower operating reliability. World-class reliability organizations employ RFCA methodologies to identify the latent root causes of preventable failures so that effective improvements can be implemented.

Introduction

A steady and disciplined root cause failure analysis (RCFA) effort is one of the pillars of an effective machinery reliability program (see Figure 16.1). The belief is that by taking actions based on RCFA's findings, future unplanned events will be reduced, and that environmental and safety performance will be improved. RCFA programs just makes good business sense.

Industry data shows that 50% to 90% of equipment failure are rooted in people and system issues. By applying RCFA methods to failures we

Robert X. Perez (ed.) Maintenance, Reliability and Troubleshooting in Rotating Machinery, (271–332) © 2022 Scrivener Publishing LLC

can garner numerous learnings about the events leading up to and immediately following the failure. We need to act on these hidden factors that result in costly failures.

What Is a Root Cause Failure Analysis?

Root cause failure analysis (RCFA) is a class of problem-solving methods aimed at identifying the root causes of problems or events. The practice of RCFA is predicated on the belief that problems are best solved by attempting to correct or eliminate root causes, as opposed to merely addressing the immediately obvious symptoms. By applying corrective measures address to root causes, it is hoped that the likelihood of problem recurrence will be minimized. However, it is recognized that complete prevention of recurrence by a single intervention is not always possible. Thus, RCFA is often considered to be an iterative process, and is frequently viewed as a tool of continuous improvement.

Furthermore, an RCFA is a formal analysis of an equipment or system failure with the intent of uncovering all physical and latent, i.e., hidden, causal factor(s), so that similar future failures may be prevented. Here are a few quick examples of latent causal factors:

- Operation of pumps in parallel causes high bearing loads leading to premature failures. The problem is not related to the pump readings but rather to how the pumps are operating. Resolution of this issue may require a new operating procedure or pump upgrade.
- Improper reciprocating compressor unloading sequence leads to connecting rod bushing overload. Under certain conditions the connecting rod bushing may not have an adequate reversing load. The compressor manufacturer may have to get involved to correct the load sequencing issue.
- Low flow pump operation due to process changes results in seal failure from excessive shaft deflection and vibration. Under normal design conditions the seal will have a long life, but under abnormal conditions the seal will fail prematurely. This problem could be resolved with a spill-back line to avoid low flow conditions.

- Oil contamination is caused by poor lubrication storage practices. New lubrication storage facilities may be required to resolve this issue.
- Oil consolidation effort leads to unwise oil replacement that results in a bearing failure. Better lubrication management practices are required to avoid this type of failure in the future.

Root Cause Failure Analysis Example #1: Ill-Advised Bearing Replacement

History

A petrochemical refinery plant was experiencing recurring failures of a cylindrical roller bearing in a critical centrifugal pump. A transition from belt-drive to direct-drive of the pump led to the unexplained bearing failure, resulting in costly downtime.

Corrective Measures

The bearing manufacturer investigated the application conditions and the failed bearings and found the solution. The switch from belt-driven to direct-driven configuration had drastically reduced the radial load on the existing cylindrical roller bearing, which had led to rollers slipping on the inner raceway. Future failures were eliminated by replacing the original bearing with a deep groove ball bearing which was designed to operate a with a lower radial load.

Comments

The physical root causes of the pump downtime were the bearing failures. The latent root cause was the transition from the belt drive to direct drive configurations without a technical review. This issue could have been avoided by discussing the design change with the manufacturer before making the conversion. The manufacturer would have surely spotted the probe and selected the proper bearing for the new load conditions.

Root Cause Failure Analysis Example #2: Reciprocating Compressor Rod Failure [1]

Background

During the repair of a large reciprocating compressor, the maintenance supervisor discovers a damaged compressor rod, requiring replacement. So, he decides to have a replacement rod manufactured in a local shop. Without consulting the original equipment manufacturer (OEM), the machine shop decides to fabricate the rod with cut threads. The OEM's experience with this compressor has led them to convert all compressor rods for compressors of this frame size to rolled threads. As a result of using cut threads, the rod fails due to fatigue in the thread area, causing extensive secondary damage inside the compressor.

Here are the chain of events leading to the failure:

- The warehouse did not stock a spare rod because it was a new compressor installation.
- The maintenance supervisor decided to have a replacement rod fabricated without drawings.
- Neither the user or the fabrication shop investigated the thread requirements.
- Because the compressor was not equipped with vibration shutdowns, it ran for a significant amount of time before it was shut down.

Physical root cause

- Installing a compressor rod with cut threads, instead of the rolled threads.

Latent root causes

- Fabricating a replacement part without engineering or OEM input.
- Not having major OEM replacement parts in stock before start-up.
- Not having vibration shutdowns on the compressor.

Comments

This example points out the differences between physical and latent root causes. Without digging into this failure, the organization may not have uncovered the latent causes of the failure. Once the analysis is complete, we need to always ask the question: Can these learnings be used to prevent similar failure on similar equipment at this facility? You can use RCFA findings to look for similar machine and organizational deficiencies.

RCFA Steps

Effective RCFA follow the following steps:

1. Define the problem.
2. Gather data/evidence.
3. Ask why and identify the causal relationships associated with the defined problem.
4. Identify which causes if removed or changed will prevent recurrence.
5. Identify effective solutions that prevent recurrence, are within your control, meet your goals and objectives and do not cause other problems.
6. Implement the recommendations.
7. Observe the recommended solutions to ensure effectiveness.

I will now discuss these seven steps in detail.

Step 1: Define the Problem

The first step in an RCFA is defining the issue that you want to investigate. It is important that your problem statement is not too broad. The clearer the problem statement, the greater the probability of determining the root cause.

Which of these two problems statements more specific?

A: Frequent pump failures in the plant are causing numerous process outages

B: The 4/25/09 Recycle Pump failure led to a 48-hour Dehydration Unit outage.

Obviously the second problem statement is more defined and manageable.

Step 2: Gather Data/Evidence

With any failure or problem, it is of the utmost importance to collect data with respect to that problem. Consider what NTSB investigators do right after an airplane crash. They comb the area for data such as the data flight recorder ("black box"), broken airplane parts, instrument readings, etc. You must do the same when you are analyzing a problem. As an analyst, you must, through interviews, brainstorm with all of the people involved to collect the other items such as:

- Failed parts
  - ✓ Gather all damaged pieces.
  - ✓ Protect fracture surfaces.
  - ✓ Collect samples of suspicious liquids, coatings, deposits, etc.

- Process data
  - ✓ Collect process data immediately. Data stored on computers tend to get archived and consolidated.
  - ✓ Assign a process person the task of collecting the data. They probably know the process better than you.
  - ✓ Don't forget to gather logbook information and operator note.

- Eyewitness testimonies
  - ✓ Interview operators as soon as possible. Stories change with time.

- Site photos.
  - ✓ Take lots of photos before anything is moved.

Identifying the Physical Root Cause of the Primary Failure

For RCFA findings to be valid, it is essential that component failure modes be correctly identified. The process of identifying analyzing failure mechanisms involves:

1. Identifying the damaged components. Be sure to secure all the damaged parts so they are not thrown away or further damaged due to exposure to the elements.

2. Determine which failures are secondary and which failure was the initial failure. The component that failed first is the primary failure, or the "physical root cause" of the failure. The goal of a condition monitoring program is to detect primary failures before they fail catastrophically. Components that were subsequently damaged are secondary failure modes. For example, if a bearing fails and causes a shaft to rub excessively, due to excessive motion, the bearing failure is the physical root cause and the shaft failure is a secondary failure.
3. Examine and analyze the primary failure by carefully cleaning the component and then looking at it under low-power magnification, 5x to 50x. Then ask the following questions: What do the failure faces look like? Are there other cracks or suspicious signs in the area of the failure? Important fracture surfaces should be photographed and preserved for reference.
4. Characterize the failed piece and the support material. Perform hardness test, dye penetrant and ultrasonic examination, lubricant analysis, alloy analysis, etc. Examine the failed part and the components around it to understand what they are. Check to see if the results agree with design conditions.
5. For critical failure investigations, conduct detailed chemical and metallurgical analyses to determine if material properties may have contributed to the failure.
6. Identifying the failure mechanism and the forces that caused the primary failure. Determining the failure mechanism can be the most challenging step in the RCFA process. *(Note: We will cover a few common component failure modes here. However, more information on analyzing and identifying failure mechanisms can be found at the end of this chapter.)* There are several ways to improve your chances of correctly identifying the primary failure mechanism:

    a. Maintain a library of component failure references.
    b. Employ an outside laboratory to assist you.
    c. Ask for help from manufacturers and vendors; they can be a wealth of failure information.

Failure modes tend to have unique visual characteristics. For example, here are some commonly encountered component failure modes along with how they can be identified:

Ductile failures, which are indicative of overloading to the point of exceeding the material's yield point. Ductile failures are characterized by necking at the point of failure (see Figure 16.2 and B1).

Fatigue failures occur whenever a material is subjected to cyclical loading and unloading for an extended period of time and the resulting stress levels are below some limiting design threshold. Beach marks, a series of nested curves resembling tide marks on a beach, indicate a fatigue crack progression through a metal component (see Figure 16.3). The first stage is a propagation of the crack over a number of cycles and is evident by beach marks. Eventually, the crack reaches a critical size, and the component fails

Figure 16.2 Ductile Failure (By Bradley Grillomn at the English Wikipedia, CC BY-SA 3.0, https://commons.wikimedia.org/w/index.php?curid=7486633).

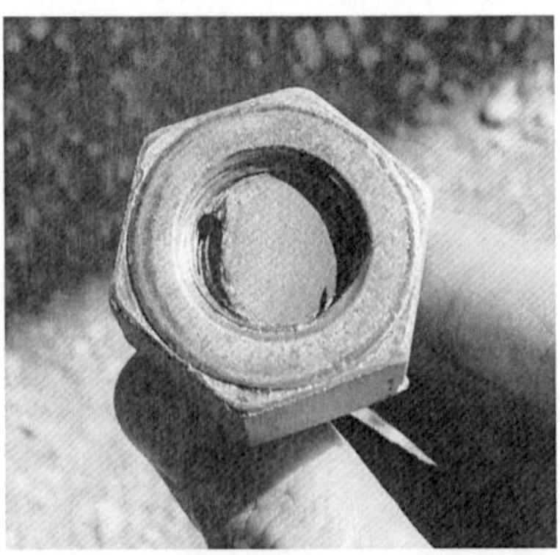

Figure 16.3 Fatigued hold down bolt caused by excessive piping vibration.

rapidly. Causal factors such as high vibration and pressure pulsation levels, and unsteady flows are often to blame.

When fatigue failures are suspected look for beach marks. By examining the beach marks closely, you may be able to determine the origin of the crack propagation. Fatigue cracks commonly original at keyways and shaft steps. Fatigue failures can be addressed by either, a) reducing the magnitude of the dynamic loading, b) increasing the endurance limit of the material, or c) increasing radii to reduce the effect of stress risers.

Fatigue Example: Fin-Fan Cooler Shaft Failures

Here is a shaft fatigue case study that explains how to analyze and correct design issues responsible for shaft fatigue failures: A group of 30 HP, 1750 rpm, fin fan air coolers experienced a total of 35 shaft failures over 4 years. All the shafts fail in a similar manner as seen in Figures 16.4 and 16.5. One of the failed shafts was sent to a metallurgical lab for a detailed analysis. Here are the lab's findings and conclusions:

1. All the shaft failures exhibited clear beach marks and striation associated with fatigue.
2. The fracture analyzed initiated at points along the corner of the shaft keyway and propagated along a plane approximately 45 degrees to the shaft axis.
3. The hardness, tensile strength, and microstructure of the shaft were consistent with the as-rolled medium carbon steels.
4. The material composition of the shaft was consistent with medium carbon steel.
5. No defects of significant surface imperfections were found on the shaft analyzed.
6. The shaft analyzed failed due to torsional fatigue.

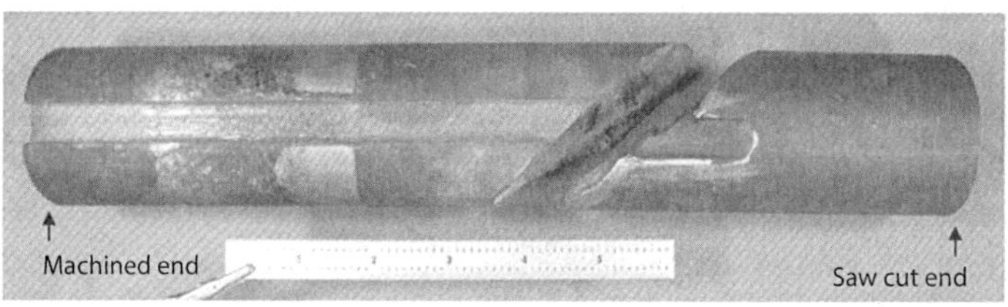

Figure 16.4 All the fan drive-shaft failure occurred at a 45-degree angle, which is indicative of a torsional fatigue failure.

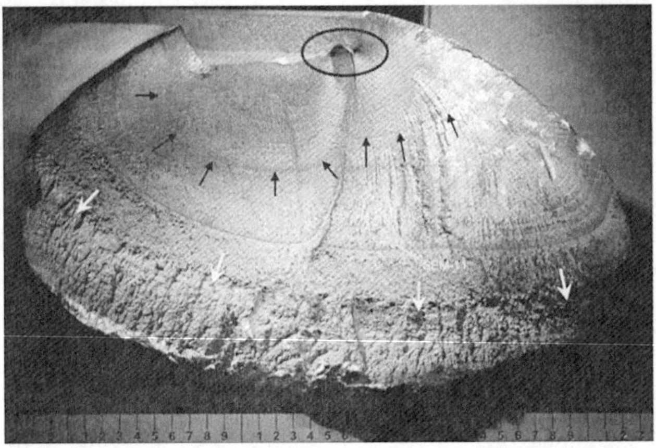

Figure 16.5 Close-up view of the fracture surface. The location of the crack origin is circled. Small black arrows show the beach marks formed as the crack progressed through the shaft. White arrows indicate the final overload region.

The final recommendations to prevent recurrence of this failure were:

1. Radius shaft keyseat to 0.0625." (The original keyseat radii were 0.006.")
2. Upgrade the drive shaft material to E4340. (The original drive shaft material was mild steel.)

The fin fan cooler manufacturer decided to upgrade the drive shaft material to 4140 instead of the E4330. This material upgrade has prevented the recurrence of torsional fatigue failures.

Fretting Wear is surface damage that occurs between two contacting surfaces experiencing minute cyclic motion. At the contact areas, lubricant is squeezed out, resulting in metal-to-metal contact (see Figure 16.6). When fretting wear occurs in a corrosive environment, both the rubbing-off of oxide films and the increased abrasiveness of the harder oxidized wear debris tend to greatly accelerate wear. When corrosion activity is clearly evident, as denoted by the rust color of the debris particles, the process is referred to as fretting corrosion.

Fretting decreases the fatigue strength of materials operating under cycling stress. This can result in fretting fatigue, whereby fatigue cracks can initiate in the fretting zone. Afterwards, the crack propagates into the material. The best way to prevent fretting failures is to minimize or eliminate relative motion between machine components through

Figure 16.6 Example of shaft fretting.

design and by ensuring component fits are correct at the time of assembly. Lubricants may be used to mitigate fretting by reducing friction and inhibiting oxidation.

When identifying the failed compnents, here are few things to keep in mind:

1. Compare the metallugical analysis with the manufacturer's latest design specifications. It is common for older machine components to have been redesigned over the years.
2. Ensure that any coatings were correctly specified and applied.
3. Ensure that the proper heat treatments were applied. Hardness testing can help determine if the components have the right material properties.
4. Inspect radii to ensure they are correct.
5. Check the current process loads to see if they are normal or have increased.
6. Check for changes in enviormental factors, such as particulates, contaminants, or temperature increases.
7. Check for any significant changes in vibration levels.
8. Examine repair records to see if similar failures have been experienced in the past.

Preserving Machine Data

To undersand machine loads, we need to inderstand how a machine is being operated and what condition it is in. We want to know: 1) Was the machine being operated as intended? 2) Were their any indications of degradation before it failed? 3) What were the possible sources of excessive loads? To answer these questions we can look at:

- Process trend data, like that seen in Figure 16.7, which shows a series of compressor surge events.
- Trends of machine parameters, such as bearings and discharge temperatures, vibration, flows, speed, etc.
- Correlations of various machine parameters versus process and enviromental conditions.
- Machine vibration analysis, i.e., frequency and waveform analysis and pressure pulsation data. Analyzing vibration data can provide clues to potential souces of dynamic loads.

Before the RCFA team can effectively begin determining the sequence of event and the root causes of the failure, they need to gather a "critical mass" of field data, operating knowledge, and history. This collection of vital information is sometimes called "front-loading elements," because it is required before any failure theories or sequence of events are developed by the investigation team. Front-loading elements include:

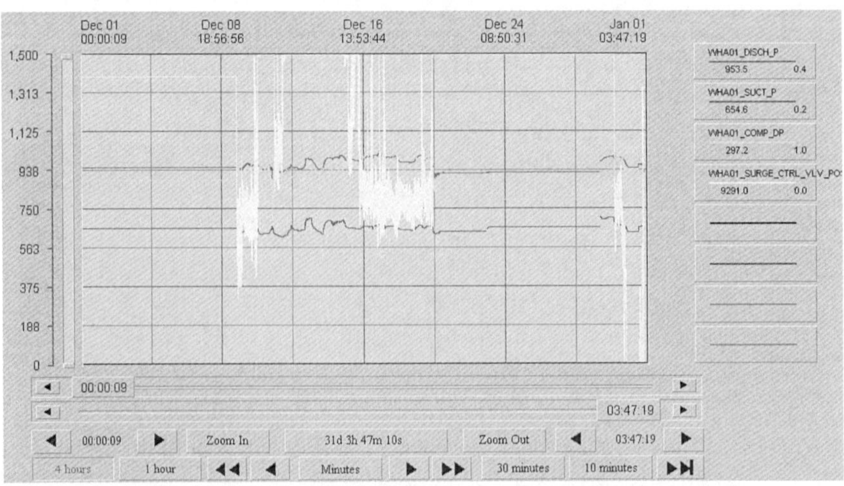

Figure 16.7 Compressor upset trend plot showing multiple surge events.

a. Equipment history and performance data, which can be generated from the work order history, the process historian (PI, DCS, etc.), or interviews with operators and craftsmen.
b. Equipment reliability strategies, which are gathered by auditing established predictive and preventative maintenance programs and determining if these programs are active and current.
c. Equipment life cycle documents, which include new purchase specifications, can sometimes reveal changes to equipment as a result of a previous learning. Installation and commissioning standards also may uncover problems.
d. Equipment operating expectations and design limitations: What are the equipment's current operating expectations? Are there any process reliability limits (pressures, flows, temperatures) set for this equipment?

Step 3: Ask Why and Identify the Causal Relationships Associated with the Defined Problem

Once you have collected the front-loading elements you can begin to ask what caused the failure.

Causal Chains

One of the keys to understanding what happened is identifying the causal chain leading to the event of interest.

Consider the following:

An operator uses a high-pressure water hose to clean off dirt, oil, and grime from a centrifugal pump (Figure 16.8). This action allows water to get into the bearing housings, which contaminates and eventually degrades

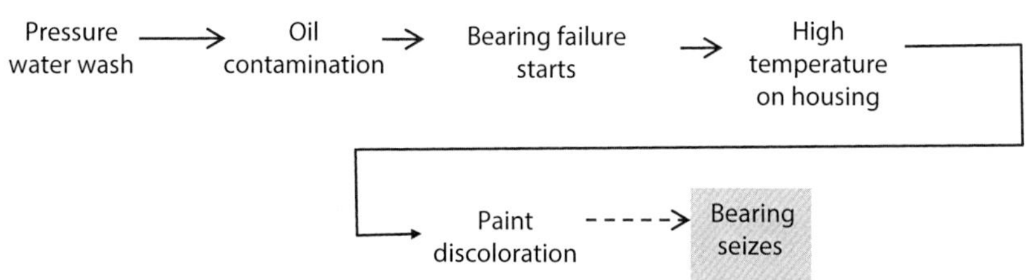

Figure 16.8 Causal chain depicting the events leading to a catastrophic bearing failure.

the lubricating oil. The degradation of the oil results in higher than normal bearing temperature, which in turn leads to an even higher bearing temperature, and so on, eventually causing the paint close to the bearing to discolor due the localized heating. In this case, an operator spots the bearing housing discoloration and shuts the pump down before catastrophic damage occurs. If the bearing degradation had not been detected, the bearing would have overheated and failed catastrophically. Note that the last arrow in the casual diagram is dashed, indicating that the sequence of events was stopped by the operator's actions.

During the pump's repair, the most obvious indication of the mechanical failure would be the damaged bearing, so it could be called the physical root cause of the failure (see causal chain explanation below). But if we stopped our investigation there, we would never understand the true root cause of the bearing damage. By moving further to the left, i.e., back in time, on the casual chain, we can peer deeper into the chain of events and identify pressure water washing as the latent root cause, i.e., hidden, root cause. Simply by looking at the bearing during disassembly, it may not be obvious why the bearing failed. It would take some detective work to discover that the bearing failed due to water ingress due to water washing.

High bearing temperature and paint discoloration fall to the right of the pressure wash event, i.e., later in time, of the physical root cause. High temperature and paint discoloration associated with high temperatures are called symptoms because they are perceptible signs of distress (see causal chain below) that are a consequence of an internal machine defect.

Bearing Failure Sequence of Events with Descriptions

Pressure Wash (Latent root cause) ==> Oil Contamination (Intermediate event) ==> Bearing starts to fail (Physical root cause) ==> High Temp (First symptom) ==> Paint Discoloration (Second symptom) ==> Bearing seizure.

To uncover latent root causes, first, discover the physical root cause of the failure and then ask yourself: *"What factor or factors could have led to this damage?"* For example, let's say that the physical cause of a fan outage was a bearing failure due to high vibration. They primary question might be "Why were there high vibrations?" To address this, we can ask the following questions:

1) Why did the bearing fail (physical root cause)?
   The bearing failed due to high vibration.
2) Why did the system experience high vibrations?
   The high vibration was the result of fouling of the fan.
3) What caused the fouling?
   The fouling was caused by abnormal process conditions.

4) What caused the abnormal process conditions?
The upset was caused by abnormally high process temperatures due to a bad thermocouple (latent root cause).

We could conclude that a bad control temperature value was the root cause of the fan bearing failure. One way to avoid this type of failure in the future would be to install multiple thermocouples that would allow the control system to continue operating with a failed thermocouple. Or, have an alarm when the temperature at the thermocouple is outside an expected range.

To maximize machinery safety and reliability, we recommend these causal chain guidelines be followed:

1. It is best to terminate the causal chain as early as possible. In the example above, it would behoove the machine owner to detect either oil contamination using oil analysis methods or bearing failure using vibration analysis methods as early as possible before bearing degradation symptoms are noted.
2. The most cost-effective means of preventing a failure, or a future occurrence, is eliminating the latent root cause. In the first example, the latent root cause is pressure washing. A procedural change is required to make sure operators are instructed to avoid pressure washing machines as well as to avoid sensitive areas of machines.
3. To avoid costly secondary damage, employ proven analysis methods to detect machine faults in the earliest stages of failure. In the first example, we can use oil analysis and vibration analysis methods to detect problems in the primary state.
4. Always strive to uncover the latent root cause of unwanted events. The latent root cause of a problem is the causal factor that, if properly addressed, will prevent a recurrence of the problem. Addressing possible causes that are not root causes will either have no effect or merely mask the real problem.

Note: Refer to Appendix A for more information on causal chains and sequence of events.

There are two popular ways to investigate mechanical failures. Because mechanical failures typically have a well-defined cause and effect relationships, they can be readily analyzed by the Why method and the cause maps.

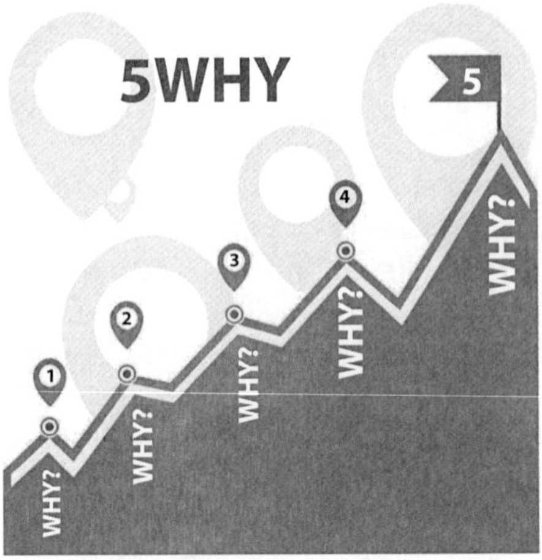

Figure 16.9 The 5 why RCFA method requires the investigator to keep asking "why" until the root cause of the failure is determined.

Five Why RCFA Example

Here is a simple RCFA example using the 5 Why method, where the analyst keeps asking "why" until he or she arrives at the event that started the chain of events that lead to the undesired condition (see Figure 16.9). In this example, we investigate a centrifugal compressor issue:

Problem: The flow from a critical centrifugal compressor dropped off suddenly, resulting in an unplanned unit outage. To better understand the event, you were asked to use the 5 Why Method to find out the root cause. You ask the following series of questions:

Why? Because the centrifugal compressor speed dropped below the rated speed. (First why.)
Why? Because the compressor's steam turbine driver could not maintain the rated compressor speed. (Second why.)
Why? Because the steam turbine is not getting enough steam into its steam chest (Third why.)
Why? The steam turbine's inlet strainer was found to be plugged. (Fourth why.)
Why? The startup inlet strainer was never replaced with a larger mesh, permanent strainer as required. (Fifth why.) **Root cause.**

The 5 Why Method is simple and straightforward. Technicians can use it to evaluate failures and determine potential causal factors.

Cause Mapping [2]

A cause map provides a visual explanation of why an incident occurred. It connects individual cause-and-effect relationships to reveal the system of causes within a failure sequence. A cause map can be very basic, and it can be extremely detailed, depending on the issue. Consider the cause map shown in Figure 16.10. You read this cause map by starting on the far left, where the problem statement is located. Then you read to the right of the page while saying "was caused by." In this example, we would say, "The problem was caused by event #1, which was caused by event #2, which was caused by event #3, and so on, until you reach the farthest left block, which describes the initiating event. In simple cause maps, the initiating event is the root cause of the problem.

Figure 16.11 shows a cause map explaining why a journal bearing failed. To understand the nature of the failure, we start the final block, which reads "High pressure amine pump was manually shut down by plant mechanic." Then the questions begin. We ask, "Why did this event happen?" The response to this question provides a cause (or causes), which is written down to the right. In our example, the pump was shut down because the mechanic noticed "Smoke was coming out of the outboard bearing housing." The next question is again, "Why did this effect happen?" The cause that was written down last becomes the effect for the next Why question. The mapping process continues to the right until all the events have been documented. The cause map is complete when all the cause-and-effect elements have been defined all the way back in time to the earliest event of interest.

"AND" gates may be required when more than one cause is required to cause an event. For example, a torsional fatigue failure requires high torsional stress, a stress riser, and a low endurance limit material to be present. In our amine pump example, wire-wooling required four factors to be present for wire-wooling to occur: a) a shaft with a chrome content greater than 1% to 1.5% content, b) a shaft speed greater than 67 ft/sec, c) a hydrocarbon environment, and d) a hard foreign particle in the bearing. When

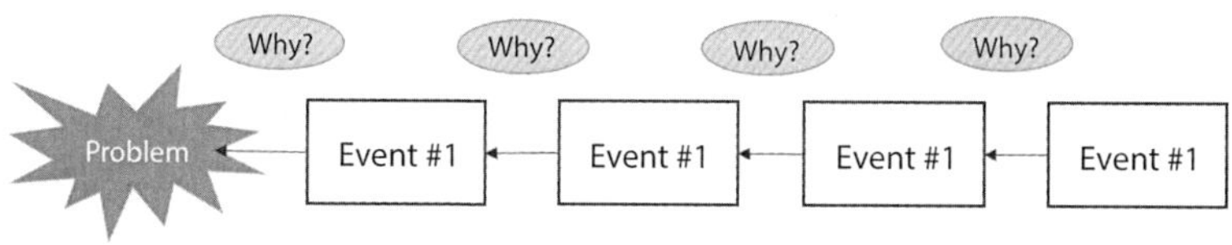

Figure 16.10 Shows a simple cause map explaining the events that led to a hypothetical problem.

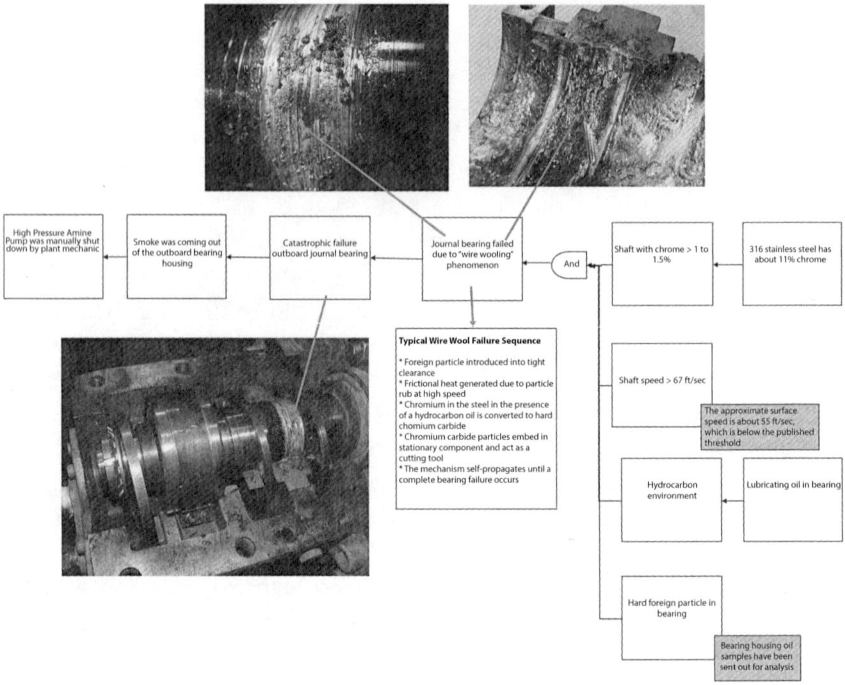

Figure 16.11 Journal failure due to wire-wooling.

an effect has more than one cause, all the causes must be verified and then documented by placing them on the cause map. While these causes are independent of each other, all are required to produce that effect. People think of cause-and-effect as a simple one-to-one relationship; an effect has a cause. In reality, every effect may have several causes.

A second causal example requiring an "AND" gate is stress corrosion cracking (SCC), which is a common problem affecting turbomachinery components such as rotor discs, steam turbine blades and compressor impellers. The following three factors must be present for SCC to occur: 1) An alloy is susceptible to stress corrosion cracking; for example, mild steels are susceptible to alkalis and stainless steels are susceptible to chlorides, 2) a corrosive environment, and 3) sufficient tensile level of stress. If any one of these causal factors are not present, then stress corrosion cracking will not occur. (Note: The mechanical design of the component, its operating environment and other factors, such as unexpected stresses, should be evaluated when determining remedies to SCC.)

An "OR" gate is used when any one of several factors can cause an event to occur. For example, an "OR" gate could be used in the case of a pump impeller erosion, where the erosion could be caused by either cavitation, particulates in the pumpage, or off-design operation. A second causal example requiring an "OR" gate would be one explaining why a high centrifugal compressor is experiencing a high thrust bearing temperature. The "OR" gate could include the following causal factors: 1) a higher than normal thrust load due to an increase in the compression ratio, 2) excessive wear in the balance piston seal, 3) possible thrust bearing pad damage, 4) a loss of lubrication flow, 5) a build-up of varnish on one or more thrust pads, or 6) a change in the lubricant properties, i.e., contamination. (Note: Readers could choose to include additional possible causal factors in the ones listed here based on their experience and knowledge of the machine.)

In this cause map failure example, we concluded that the combination of a shaft with greater that 1% to 1.5% chrome and a shaft journal speed greater than 67 ft/sec were the physical root cause of this failures. The latent root cause was the lack of oversight during the pump repair that led to the selection of the wrong shaft metallurgy.

Cause Map Example #2

The next cause map example documents the catastrophic failure of a critical process compressor. Starting from the far left (see Figure 16.12), we can see that production goals and maintenance goals were impacted by this failure. Next, we can see the reason for these impacts was the failure of coupling between two of the compressor bodies. This event was caused because the #2 compressor completely lost its axial position due to a failure of its thrust bearing and lack of a functioning axial position trip system. As we follow the cause map from left to right we find that there were several physical root causes of the failure:

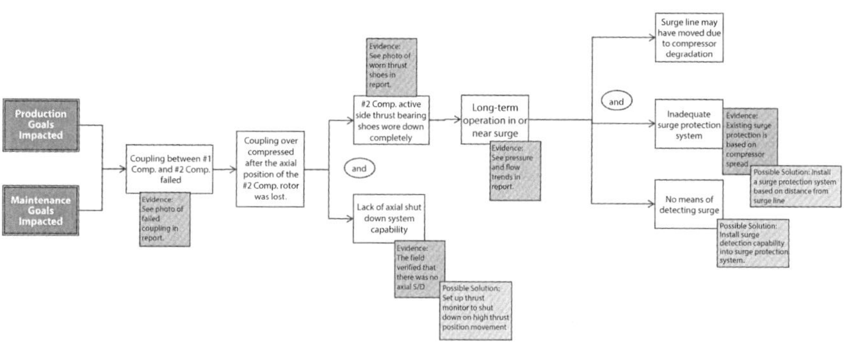

Figure 16.12 Cause map of catastrophic compressor failure.

1. Compressor degradation
2. Inadequate surge control
3. No surge detection.

The latent root cause, which are not included in the cause map, was concluded to be the general negligence of the maintenance and operating departments that allowed the surge and axial position protection systems to fall into disrepair. These critical protection systems need to be maintained and tested on a regular basis to ensure their integrity.

The beauty of this cause map is that it's easy to read and understand. It also contains information that supports the various findings and possible solutions. Management will be able to quicky digest the findings and conclusion of the investigation team.

Single Root Cause versus Multiple Causes

The "old" way of thinking was that every failure or event had only one root cause. Any additional issues were often called contributing factors. Now, cause mapping adopts the thinking that failures or events can result from multiple causes. Cause mapping doesn't distinguish between root causes and contributing factors. This distinction tends to open up discussions between team members and generate more potential causes.

Allowing multiple causes on your cause map has several advantages. Generating multiple root causes can:

- Reduce conflicts in the team because all causes are considered.
- Improve your chances of finding subtle factors.
- Encourage group participation by allowing all points of view to be considered.

Cause Mapping Steps

The three main steps involved in the cause mapping method are:

1. Define the issue by its impact to facility goals by asking: What was or were the consequence(s) of the problem? Problems might result in one or more of the following consequences: 1) a production goal is impacted, 2) a maintenance budget goal is impacted, 3) an environmental goal is impacted, or

4) a safety goal is impacted. The complete problem statement should include a title, the date the event occurred, the location of the event, and any other details, such as which facility goal or goals were impacted, to help the analysis team understand the problem.

2. Analyze the causes that led to the problem with a visual cause map by asking: Why did it happen?
3. Prevent or mitigate any negative impact to the goals by selecting the most effective solutions by asking: How can this failure be prevented from recurring?

Along the way, carefully document your analysis steps and findings. When you are done, your analysis report should have a clear problem statement, a cause map explaining the chain of events, and a list of action items that you believe will eliminate the root cause(s).

4. Identify the causes that will break the causal chain and prevent recurrence of the failure (see Figure 16.13).

Once you have determined the chain of events that led to the failure of interest, it is easy to visualize what happened and then determine the best way to break the event chain. Let's go back to the reciprocating compressor rod failure example mentioned at the beginning of this chapter. The following sequence of events led to the rod failure:

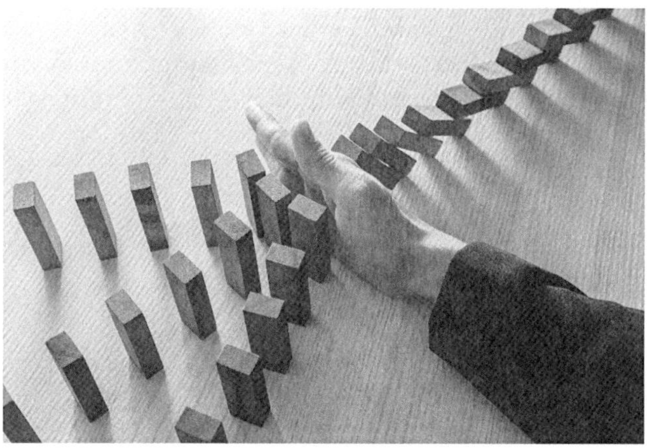

Figure 16.13 Breaking the chain of events.

1. No spare piston rods were ordered.
2. Rod scoring occurred on one of the rods requiring it to be replaced.
3. Decision made to make a replacement rod instead of ordering one from the manufacturer.
4. A local score made a rod with cut threads instead of the recommended rolled threads.
5. The fabricated replacement rod fails.
6. Excessive secondary damage after the rod failure was experienced due to a lack of vibration shutdown system.

This failure sequence could have been stopped by either:

- Stocking spare compressor rods in the warehouse, or purchasing a spare rod from the OEM.
- Requiring the shop from fabricating critical spare parts without engineering support.
- Installing a vibration shutdown system to prevent secondary damage.

The suggested sequence to avoid repeat failure is: First, you must fully understand the event chain and then identify its root cause or root causes. Finally, you can implement design improvements, procedural improvements, training programs, etc., aimed at eliminating unwanted originating events and latent factors that address the root cause(s) of the failure.

Identifying the correct root cause or root causes if crucial to preventing recurrence of a similar failure event. You will improve your chances of identifying all the root causes by gathering lots of data and information and generating a lot of ideas on the front end of the investigation. Try to avoid focusing on root causes too early in the investigation.

Here's a partial list of categories to consider when selecting your root causes:

1. System, equipment or component issue:
   a. Design error
   b. Misapplication
   c. Defective part
   d. Improper machine repair

2. Staff/People/Training issue
   a. Inadequate skills

   b. Lack of training
   c. Moral issues

3. Rules/Policies/Procedural Issue
   a. Lack of procedures
   b. Inconsistent application of policies
   c. Outdated procedures

4. Organizational issues: A system, process, or policy that people use to make decisions or perform work is faulty.
   a. Example 1: If during a machine repair there is no formal procedure to transfer information from one shift to the next, miscommunication could lead to mistakes
   b. Example 2: If there are not formal start-up procedures for operators, improper start-ups could lead to failures.

5. Publish a final RCFA Report.
   The ultimate goal of an RCFA investigation is to publish a comprehensive report for management to review and act on. The report should include:

a. *What happened?* A discussion of the problem that was investigated and how facility goals were impacted. The problem statement should include a title, the date the event occurred, the location of the event, and any other details that will help the team analyze the problem. For example, you could write, "The East Plant, #2 Reformer Compressor failed catastrophically on May 16th, 2020. The failure resulted in a 2-week compressor outage and resulted in a $250,000 repair cost.

b. *Who was involved?* Include a listing of the team members in the report so that management can determine if sufficient resources were provided to investigate the problem.

   Your failure investigation team should have members having broad backgrounds and experiences. The key to finding the root cause or root causes is having a team with a wide range of experience that fully understands the process and the machine and is well versed in the way the organization functions or is supposed to function. For typical machinery failure investigations, it makes sense to form teams consisting of a rotating machinery professional (team lead) along with a mechanic, operator, and perhaps even a process engineer. For more complex failure events, you may wish to include a controls specialist, safety professional and a production supervisor. Each team member provides a unique perspective of the failure event. It also makes sense to draw team members from outside the operating area where that the failure occurred to prevent bias.

c. *Why did failure occur?* A description of the root cause(s) uncovered by the investigation team. Include a cause map or 5 Why sequence of events to visually explain what happened. Any additional technical discussion required to explain the sequence of events should also be included.
d. *How can this failure be prevented from recurring?* A critical step in the RCFA process is determining which prevention or mitigation recommendations to present to management. Here are some guidelines to ensure your action items are effective:

- *They address root cause(s):* Only action items that address the root causes should be forwarded to management. Here are two recommendations that illustrate the differences between acceptable and unacceptable action items:

  Action items, example #1: Let's assume that the root cause of repetitive mechanical seal failures has been determined to be flashing at the seal faces due to a lack of cooling. A key recommendation might be to install an API plan 23 to reduce the fluid temperature inside the seal cavity and across the seal faces. However, it would not make sense for the same investigation team to recommend a larger motor driver for the pump. While someone in the investigation team might want the pump to handle more flow, this doesn't address the identified root cause and should be justified on its own merits.

  Action items, example #2: Let's assume that during a centrifugal compressor/steam turbine train outage, cracking was discovered at various locations in the blade roots of the steam turbine driver. A third-party metallurgical lab confirmed the cracking was related to stress corrosion cracking (SCC). They also noted that the three prerequisite factors required for SCC are:

  - The alloy is susceptible to stress corrosion cracking.

- The stress intensity factor is above the threshold value.
- A corrosive environment is present.

Based on the metallurgical lab's recommendations, the RCFA team recommended: a) upgrading the turbine component metallurgy, b) modifying the blade root design to reduce peak stress levels, and c) analyzing the steam composition to identify if any corrosive agents may be present. Since these three action items address the three prerequisites for SCC to occur, they are all appropriate action items for the team to consider for inclusion in the final report.

In contrast to these recommendations that directly address the root cause of the cracking, it would not make sense to recommend a control upgrade for the compressor/steam turbine train. Even though the controls may be found to be antiquated, they had no effect on the chain of events that led to the cracking. A control upgrade project would have to be justified on its own merits.

- *They contained a limited in number and costs of action items:* Only the action items required to resolve the root cause(s) should be listed. Any items that direct capital, maintenance, or engineering dollars should be challenged for their cost and probability of success. In my experience, less is more. You have a better chance of implementing three recommendations than 30.
- *They are measurable and trackable:* Action items should be clear and concise to yield a definite project scope, along with qualitative expectations and measures.
- *They are realistic and economically beneficial:* Every potential solution should have a positive economic benefit that can be quantified and defined.

It's a good idea that your team asks one or more individuals that were not involved in the investigation to read the final report before it is formally issued. An unbiased third party can provide a cold eye review and evaluate your findings and recommendations from a fresh perspective. Independent reviewers may be able to find gaps or errors in the discussion and even provide recommendations missed by the team.

6. Implement the recommendations.
   The only way to realize reliability improvements is by implementing the recommendations that are included in the final

RCFA report. What good is it if the reliability group generates 100 recommendations and none are implemented?

World-class maintenance organizations have a centralized recommendation database to track all department recommendations. The database should contain recommendations records that include a description of individual action items. Each record should include the following details:

- Background information, i.e., RCFA report or other event details that explain the reason for the recommendation.
- Technical details and requirements.
- A history of what work has been done so far, such as field investigations, engineering assessments and designs, quotations, drawings, etc.
- Economic justification(s)
- Status, i.e., active, inactive, complete, cancelled.
- Owner(s)

A periodic review of the database is recommended to ensure its integrity and alignment with current site goals. Perhaps quarterly or annual review with area and plant management of open recommendations makes sense. The review team should concentrate on past due recommendations, out of date recommendations, and any other details that the team would like to review.

7. Observe, or follow, the recommended solutions to ensure effectiveness.
   We are all busy and rarely have time to review our past work, but it is vital to go back and evaluate the effectiveness of our implemented recommendations. As we review effectiveness machine modifications to problematic equipment, we need to ask:

   a) Have any repeat failures occurred?
   b) Is the MTBR (Mean Time Between Repairs) actually improving?
   c) Were the expected results achieved? In other words, is the post-modification failure rate acceptable?

If action items are in place and no improvements are seen, we need to determine if the original RCFA should be reviewed to see if new data is available or if an additional solution should be considered. If new data or information is available, perhaps the team needs to revisit the problem to see if there is a latent root cause that has not been identified.

The step of reviewing the effectiveness of recommendations is part of the continuous improvement process.

Inhibitors to Effective Problem Solving [3]

RCFA methodology is complex, and the required investigation effort can be easily derailed if the team is not careful and loses focus. Here are some common traps that prevent RCFA investigation teams from correctly determining the root cause of a failure:

Problem described incorrectly: The problem must be clearly and thoroughly described and must be narrow enough in scope for the analysis team to handle.

Problem-solving effort expedited: Steps in the problem-solving process should not be skipped in hopes of obtaining a quick solution.

Poor team participation: Limited or spotty participation by team members may lead to limited consideration of potential root causes, and therefore limit their probability of success.

No logical thought process: A team that lacks a disciplined system to prioritize, analyze, and review data is likely to flounder and require frequent kick-starts.

Potential cause misidentified as a root cause: If the problem that has been previously addressed recurs, the root cause has not been identified. The team should not have concluded the failure investigation in haste.

Permanent corrective actions not implemented: The task is not to identify the root cause, nor is it to put a Band-Aid on the problem; it is to initiate an effective, long-term corrective action that will prevent the recurrence of the problem.

Source: FORD, Team Oriented Problem Solving

When Is a Root Cause Failure Analysis Justified?

Time is Money.

The RCFA method is one of the most powerful tools of a reliability program, but it is also one of the most time consuming and hence costly endeavors. Since time is money, RCFA's must be administered judiciously.

Like all decisions made in a competitive environment, the decision of when to conduct an RCFA must be based on economics. Since not all failures are created equal, not all failures warrant an analysis. The decision to conduct an RCFA should be based on two questions:

1. What is the consequence of the failure? Obviously as the failure consequence rises the more critical is the need for analysis.
2. What is the failure frequency? The higher the failure rate, the more attention the piece of equipment should get.

If we denote the consequence as C and the rate of failure F, we can define the risk (R) of failure as the product of consequence and the failure rate, i.e.:

$$\text{Risk(R)} = \text{C x F}$$

So, if a pump that costs $20,000 to repair fails 4 times a year, this represents an annualized risk of $80,000 per year, i.e., $20,000 x 4. Next, we will construct a risk matrix based on a broad range of consequences and failure frequencies (see Figure 16.14 below):

Notice that the risk levels, i.e., consequence times frequency, over $1,000,000 per year are highlighted in a rose color, risk levels of $100,000 per year are highlighted in yellow, and that risk levels of $10,000 per year are highlighted in green. (Note that risk levels $1,000 and below are grayed out because these risk levels are of little concern.) In general, we can say

		Annual Failure Frequency				
		0.01	0.1	1	10	100
Consequence per Event	$ 10,000,000	$100,000	$1,000,000	$10,000,000	$100,000,000	$1,000,000,000
	$ 1,000,000	$10,000	$100,000	$1,000,000	$10,000,000	$100,000,000
	$ 100,000	$1,000	$10,000	$100,000	$1,000,000	$10,000,000
	$ 10,000	$100	$1,000	$10,000	$100,000	$1,000,000
	$ 1,000	$10	$100	$1,000	$10,000	$100,000
	$ 100	$1	$10	$100	$1,000	$10,000
	$ 10	$0	$1	$10	$100	$1,000

Figure 16.14 Consequence versus failure frequency.

that high-risk issues are in the upper right-hand corner of risk maps and low-risk issues are in the lower left-hand corner of risk maps.

You can consider these annualized risk levels to be potential revenue that can be recouped if the root cause of an issue falling in one of these cells is found and the required recommendations are implemented. With that being said, we can consider all RCFA as potential projects—and like any projects we must consider their cost to benefit ratio.

Suppose we have an event that is occurring 10 times a year with a consequence of $10,000. This represents an annualized risk of $100,000. If the total cost of the recommendations and RCFA is $20,000, we can pay this project back in about 73 days! On the other hand, if you have a $100,000 consequence that occurs once every 100 years, this only represents a risk of $1,000 per year. It wouldn't make much sense to assign a multidisciplinary analysis team to analyze this failure.

The takeaway here is that it is the annualized risk is a better means of determining the level of analysis justified—not the consequence level. However, it is possible to also set a maximum allowable loss trigger for a failure investigation. For example, a site may choose to investigate all events resulting in losses of $1,000,000 or more, regardless of their frequency. Together, the risk map and the maximum allowable loss methods can provide clear RCFA guidance.

Most production companies take this basic risk-based approach and develop their own risk matrix as show in Figure 16.15 below:

Each organization must define what high-, medium-, and low-consequence events are, along with low, medium, and high failure frequencies. For example, a company might choose to define high-, medium-, and low-consequence events as those resulting in losses of more than $250,000, $100,000 to $249,999, $25,000 to $99,999, respectively. High, medium, and low failure frequencies could be defined as more than two failures a year, more than one failure every 2 years, less than one failure every 2 years,

High Consequence	A	A	A
Medium Consequence	B	B	A
Low Consequence	C	C	B
	Low Failure Frequency	Medium Failure Frequency	High Failure Frequency

Figure 16.15 Typical risk matrix.

respectively. So, a $150,000 failure occurring every year would fall in the "B" risk level category, but a $150,000 failure occurring twice or more a year would fall in the "A" risk level category.

Non-monetary events such as environmental releases and safety events may also be included in your risk matrix. An example of a high environmental event could be a release of 10 barrels or more of any hazardous fluid from the process. A high-risk safety event could be any loss due to an accident.

Notice in the risk map above that even high-consequence events occurring at lower frequencies are considered "A" risk levels. This is because this risk map incorporates a maximum allowable loss trigger regardless of event frequency for high-consequence events. This is management's way of stating that all major events are unacceptable at any frequency.

RCFA Levels

Now that we have defined A, B, and C risk level events, we need to determine what level of analysis is warranted at each of these levels. Here are some general guidelines for assigning when and who analyzes A, B, and C level events.

Level A Issues: Failure events falling in the "A" region of the risk matrix justify the highest level of analysis. This usually means that a multidisciplinary team be used to mine the data and procure the physical evidence in order to determine the most probable cause or causes of the failure. This usually means a team of three to five technical specialists, often composed of a machinery engineer or technician, vibration tech, process engineer, operators, etc., should be formed to leave no stone unturned during the investigation. The team may take a few days or weeks to finalize their findings and recommendations.

Level B Issues: Failure events falling in the "B" region of the risk matrix are considered costly but do not justify a "full-blown" team approach. Level B analyses are typically conducted by highly skilled machinery technicians, working with process support. This type of analysis rarely takes more than one week for an individual to finalize.

Level C Issues: Failure events falling in the "C" region of the risk matrix can be considered your "bread and butter" investigations. Even though they represent low-consequence events, they typically represent the largest number of RCFAs. So, based on their sheer number, they can represent the greatest potential for savings for a plant. Level C RCFAs are often conducted by mechanics using the "5 Why" or similar methods. A systematic use of Level C RCFAs can lead to a dramatic and rapid reduction of repeat failures.

Closing Thoughts

The root cause failure analysis effort of a plant should be considered one of the main pillars of machinery reliability. I have personally seen formal failure investigations solve the most perplexing and costly problems. But to ensure success, participants must be given the proper training and the time to uncover latent root causes.

The payoff is worth the investment. Consistent administration of these methods can empower your organization to whittle down early and frequent machinery failures until only predictable end-of-life failures are experienced.

Appendix A

No Magic Allowed

Identifying Sequence of Events and Causal Chains

Determining the sequence of events that precedes a failure is an essential step in understanding why the failure occurred and how it can be prevented from recurring. However, care must be taken to ensure the events are actually related to one another and ordered in the proper sequence. Machine knowledge and experience are required to determine if an observed sequence of events actually consists of interrelated events or not. An ordered event sequence with clear cause-and-effect relationships starting from an initial event all the way to the final event of interest is referred to as a causal chain. Remember that a causal chain can be described as a sequence of events, but not every observed sequence of events can be called a causal chain.

You can think of a failure sequence as a series of dominos lined up in a row (see Figure A.1) that fall in succession after the first domino falls over. An initiating domino topples the domino adjacent to it and then the second domino topples the one adjacent to it and so on until the final domino in the line falls. If we fully understand the sequence of events that transpired, then we can identify the initiating event, which we call the root

Figure A.1 Think of the sequence of events leadings to a machinery failure as a series of dominos in row that fall in succession after the first one falls.

cause, and figure out ways to stop the sequence of events from propagating once the failure sequence has started.

I had a boss who used to say, "no magic is allowed," whenever we were trying to understand a failure sequence. What he meant was that for the sequence of events leading to the failure to be considered plausible every link in the event chain must obey the laws of physics, fall into a logical cause-and-effect order, and be likely to occur. If any of the events in the sequence is not permitted by the laws of science or the events don't fall into a logical order, then the proposed sequence probably didn't happen. Furthermore, if the probability of an event occurring is extremely unlikely or has never been seen or reported by someone in the industry, then it most likely did not occur. Probable events are those events that have a non-zero probability of occurring and regularly occur in the class of machine you are analyzing. Therefore, when settling on a sequence of events leading to a failure, you should begin by selecting a provisional causal chain that is permitted by the laws of science, which is arranged in a logical order, and is likely to have occurred based on the available evidence. If the investigation fails to uncover the root cause after your first attempt, you probably need to gather more evidence and then postulate a different failure scenario.

Let's consider some physical observations that obey the laws of physics:

1. Flow goes from high to low pressure.
2. Heat flows from hot to cold regions.

3. Gases and liquids defuse from high density to low density.
4. Contaminated fluids, such as lubricants, remain contaminated, unless they are filtered, treated, distilled, etc.
5. Damaged metal surfaces never heal themselves; A failed bearing cannot heal itself magically.
6. Efficiency only gets worse with time.

Notice that if you flip these statements around, they are no longer plausible. For example, flow cannot move from a lower pressure region to a higher pressure region unless work is performed in the fluid. Similarly, pump or compressor efficiency cannot improve over time. As internal seals and flow surfaces wear or corrode, pump or compressor efficiency usually degrades gradually until it can no longer perform its required function. There have been rare cases where build-up inside a machine has actually improved efficiency, but these are rare events that should never be considered the norm. This short list of physics-based laws illustrates a few fundamental laws that all rotating machinery obey.

Another characteristic of a plausible event chain is that all the events occur in a logical sequence. A simple example of a logical sequence is as follows: Seal Fails ==> product is released. If would not make sense to say the leaking product caused the seal to fail. Here are a few examples of logical event sequences:

1. Valve closes ==> Flow drops to zero.
2. Relief valve opens ==> System pressure drops.
3. Compressor flow drops below surge line ==> Surge occurs.
4. Loss of seal flush ==> Seal fails.

Most event sequences only make sense in one direction. The laws of physics often dictate the logical order of a series of events.

Now, let's discuss rare versus common events. I'll define a rare event as an event that has never been documented by your company or has only been seen a few times over the entire industry. Readers can tweak this definition to suit themselves. Here is a list of rotating machinery events that may be postulated to have occurred. (Note that some of the events are real and some are purely hypothetical.) I have rated each event in a range from common to extremely rare occurrences:

- Mechanical seal failure (common)
- Bearing failure (common)
- Oil contamination (common)

- High vibration (common)
- A fail-open valve fails loses its signal or power and gets stuck in the closed position (rare)
- An open gate valve vibrates closed (rare)
- Failed bearing heals itself (extremely rare, i.e., essentially impossible)
- Flow meter error due to plugged impulse line (rare)

Readers may disagree with some of my ratings. However, the point here is to realize that some events are rarer than others and should be carefully considered when including them in your chain of events. A detailed examination of the field data or failed parts will ultimately determine if a particular failure actually occurred.

5-Why Method of Investigation

The 5-why method is one technique that can be used to explore the various cause and effect relationships that you believe were involved in a failure sequence. The goal of this technique is to determine the root cause of a failure by repeating the question "Why?" Each answer leads the investigator to the next question. In most cases, after asking "why" approximately five times, you will arrive at the initiating event. The "five" in the name comes from the observation that usually "why" has to be asked five times to determine the root cause of the problem. Here is a simple example to illustrate how the 5-why method can be used to both develop the sequence of events and elucidate the initiating event in the event chain.

Let's say that a product fire occurred near a centrifugal pump. Eyewitnesses indicated that they believed the cause of the fire was a massive leak in the pump's mechanical seal. You are asked to perform a 5-why analysis of the event, so you begin by asking the first "why". During the disassembly of the pump, you have found that the mechanical seal was completely destroyed due to shaft vibration and that the rolling element bearing next to the mechanical seal had completely failed and was completely discolored due to experiencing high temperatures. You also discovered that the bearing housing seals were completely worn out. An inspection of the bearing housing oil revealed signs of water and oil discoloration. After considering all the available data and observation, you arrive at the following chain of events:

Why #1: Why did the fire occur? The fire occurred due to a hot bearing igniting the flammable product leaking from the pump.

Why #2: Why did the massive fluid leak occur? The seal leak occurred due the fact that the seal was destroyed due to excessive shaft vibration.

Why #3: What caused the excessive shaft vibration? The shaft vibration was caused by the failure of the bearing next to the mechanical seal.

Why #4: What caused the bearing to fail? The bearing failed due to contaminated lubricating oil inside the bearing housing.

Why #5: What causes the oil contamination? Worn out bearing housing seals allowed contaminants to enter the bearing housing.

Using all the available failure observation and the 5-why analysis tool, we arrive at the following causal chain:

Bearing housing seals wear out oil gets contaminated bearing fails seals fail due to vibration product is released hot bearing ignites product fire.

While we cannot state definitively this is what actually happened, we can say this is the most probable chain of events that led to the fire. Until new information contradicts this postulated sequence, it can be regarded as the best guess at explaining what happened.

Based on this most likely sequence of events, we need to think about practical ways to break the chain of events leading to final failure. Some possible proactive ways of preventing recurrence of a similar failure might be to:

1. Decrease the vibration inspection interval to identify bearing failure early.
2. Include vibration analysis methods that have a high probability of detecting the early stages of bearing failure.
3. Replace bearing housing oil on a time basis, i.e., monthly or quarterly.
4. Perform an oil analysis of the bearing housing oil on a regular basis.

In my opinion, if this is a small to medium-size pump, the first two recommendations make the most sense. A global recommendation might be to implement the first two recommendation on all pumps handling flammable products. Additionally, periodic pump inspections on pumps handing flammable products should include bearing temperature measurements and visual inspection of the bearing sight glasses for the detection of oil contamination.

Advice on Failure Sequences

Complex machinery failures usually require a lot of research and head scratching to correctly identify their root cause (or causes). As I have stated earlier, a successful failure analysis investigation knows the construction of a probable causal chain must be based on sound engineering principles, machine knowledge, and logic. Once you understand the sequence of events that led to the failure, you are ready to think about solutions and preventative measures.

Here's some practical advice that will improve your chances of identifying the most likely sequence of events leading to the machinery failure of interest:

1. Every link in the event chain must obey the laws of physics. Magic is not allowed.
2. Events must fall in a logical order.
3. Every link in the chain must be likely to occur, based on historical data and industrial experience.
4. Evidence must back up every event in the proposed sequence. Here are few examples of how evidence can either support or refute a theory:
    a. If you have postulated that a seal failure was caused by a loss of flush, then you should see signs of heat checking on the seal faces. If heat checking is not found on the seal faces, then the seal failed for some other reason.
    b. If you have postulated that oil contamination caused a bearing failure, then an oil analysis should show signs of contamination. If an analysis of the oil does not show contamination, you need to find another reason for the bearing failure.
    c. If you have postulated that multiple surge events led to a compressor's thrust bearing failure, then the signs of surge events should be found in the recent process flow data. If there are no signs of surge in the process data history, you need to explain why the thrust bearing failure occurred in some other way.

It often helps to get a second and even third opinion before finalizing your most likely chain of events. Fellow machinery specialists and original equipment manufacturers are always helpful in offering opinions based on a cold eye review of the evidence and theories.

The only way to avoid similar machinery failures in the future is by implementing the recommendations aimed at either eliminating the root cause of a failure or stopping the progression of events leading to a failure. Experience has shown that it is always best to eliminate the root cause or causes of a failure. The ultimate test for proving if you have found the true root cause is that, once it is addressed, similar failures will either be completely eliminated or the time between their occurrences will be improved dramatically. However, there will be times when one or more safeguards, such as vibration shutdowns, secondary mechanical seals, automatic isolation valves, etc., may have to be used to halt a failure progression in a situation where the root cause cannot be eliminated. Although these intermediate safeguards may not be considered the most ideal solution, they can be implemented to reduce the severity of similar failures.

Appendix B

Analyzing Component Failure Mechanisms

The purpose of this appendix is to provide those new to failure analysis a starting point for identifying component failures. This short listing of failure modes is by no means comprehensive. Hopefully, by studying this appendix the reader will begin to become familiar with the terminology and scope of mechanical failures. With time, those involved in failure

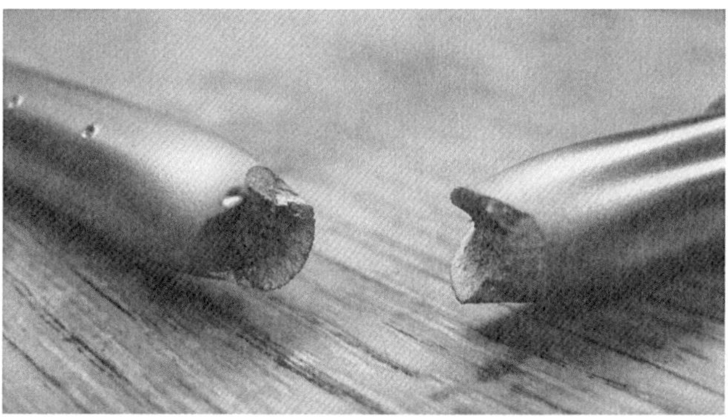

Figure B.1 Ductile failure (By BradleyGrillo at the English Wikipedia, CC BY-SA 3.0, https://commons.wikimedia.org/w/index.php?curid=7486633).

investigations should build a library of texts covering the nature and root cause of mechanical failures.

This list covers:

1. Mechanical Failures
2. Common Hydrodynamic Bearing Failure Modes
3. Analyzing Rolling Element Bearing Failures
4. Tips for Analyzing Mechanical Seal Failures

Let's begin by covering the five general main categories of mechanical failures:

1. Overload: A failure caused by applying a load large enough to exceed the material yield point in a component and cause a part to deform or fracture. The yield point, in mechanical engineering, is the load where a solid material that is being stretched begins to flow, or change shape permanently, divided by its original cross-sectional area (Stress=Load/Cross Sectional Area). The yield point, also called the elastic limit, marks the end of elastic behavior and the beginning of plastic behavior. When stresses less than the yield point are removed, the material returns to its original shape. Any increase in the stress beyond the yield point causes greater permanent deformation and eventually fracture. See deformation and flow in the image below.
2. Fatigue: A failure caused by a significant fluctuating load over a time. Failure surfaces related to fatigue tend to have beach marks, which are a series of nested curves, resembling tide marks on a beach (see Figure B.2). The advancement of a fatigue crack in a stressed component is usually caused by changes in environment and load.
3. Corrosion-influenced fatigue: Corrosion substantially reduces the fatigue strength of most metals and eventually causes failure at relatively light loads.
4. Corrosion: A failure caused a loss of material due to any number of corrosion mechanisms.
5. Wear: A variety of mechanisms result in loss of material when mechanical components rub against one another.

Figure B.2 Fatigued hold down bolt caused by excessive piping vibration.

Common Mechanical Failure Modes

Foreign Object Damage (FOD)

FOD is a term frequently used to describe damage along a flow path, such as dents, nicks, and partial fractures resulting from the ingestion of debris. These FOD-induced defects can lead to poor performance, but more concerning is unanticipated failure via fatigue cracking from stress concentrations on gas turbine blading.

Stress Corrosion Cracking

Stress corrosion cracking (SCC) is a failure that occurs when three factors are present: a tensile stress, a susceptible material, and a corrosive environment. Similar to fatigue, these cracks frequently initiate at the surface at stress levels significantly below material capabilities. Because laser peening induces compressive residual stress at the surface of a part, the potential for SCC can be drastically reduced and even eliminated.

Figure B.3 Impeller erosion.

Erosion

Erosion is the gradual, layer-by-layer destruction of the surface of metallic materials by mechanical action or electrical discharges (electroerosion). Erosion is typically seen on components in the flow path of abrasive flowing media. Coatings, surface hardening, and material upgrades may be used to prevent slow down the effects of erosion (see Figure B.3).

Cavitation

Cavitation is the formation of bubbles or cavities in liquid, developed in areas of relatively low pressure around a pump impeller. When the bubbles implode or collapse, they create intense shockwaves inside the pump, which eventually damage the impeller. If signs of cavitation are seen, the pump conditions should be reviewed to see if there's a fundamental problem. Talk to the pump manufacturer to see if they have a design solution to eliminate cavitation.

Hydrogen Embrittlement

Hydrogen Embrittlement occurs when metals become brittle as a result of the introduction and diffusion of hydrogen into the material. The degree of embrittlement is influenced both by the amount of hydrogen absorbed and the microstructure of the material. High strength grain microstructures have an increased susceptibility to embrittlement. The phenomenon usually becomes a safety concern when it leads to cracking, which occurs whenever sufficient stress is applied to a hydrogen-embrittled object. Such stress states can be caused both by the presence of residual stresses, associated fabrication operations such as forming and welding, and applied service stresses. The severity of hydrogen embrittlement is a function of temperature. Most metals are relatively immune to hydrogen embrittlement, above approximately 150°C (300 °F).

Galling

Galling is a form of wear caused by adhesion between sliding surfaces. When a material galls, some of it is pulled with the contacting surface, especially if there is a large amount of force compressing the surfaces together. Galling is caused by a combination of friction and adhesion between the surfaces, followed by slipping and tearing of crystal structure beneath the surface. This will generally leave some material stuck or even friction welded to the adjacent surface, whereas the galled material may appear gouged with balled-up or torn lumps of material stuck to its surface. Galling is most commonly found in metal surfaces that are in sliding contact with each other and is especially common where there is inadequate lubrication between the surfaces.

Fretting

Fretting wear is surface damage that occurs between two contacting surfaces experiencing minute cyclic motion. At the contact areas, lubricant is squeezed out, resulting in metal-to-metal contact. Because the low amplitude motion does not permit the contact area to be relubricated, serious localized wear can occur. This type of wear further promotes two-body abrasion, adhesion and/or fretting fatigue wear (see Figure B.4).

When fretting wear occurs in a corrosive environment, both the rubbing-off of oxide films and the increased abrasiveness of the harder oxidized wear, debris tend to greatly accelerate wear. When corrosion activity is clearly evident, as denoted by the rust color of the debris particles, the process is referred to as fretting corrosion.

Figure B.4 Shaft fretting.

Hot Corrosion (Gas Turbines) [4]

Hot corrosion is defined as accelerated corrosion, resulting from the presences of salt contaminates, such as Na_2SO_4, NaCl, and V_2O_5, that combine to form molten deposits, which damage protective surface oxides.
During combustion in the gas turbine, sulfur from the fuel reacts with sodium chloride from ingested air at elevated temperatures to form sodium sulfate. The sodium sulfate then deposits on the hot-section components, such as nozzle guide vanes and rotor blades, resulting in accelerated oxidation (or sulfidation) attack (see Figures B.5 and B.6). This is commonly referred to as "hot corrosion."

High-temperature alloys that suffered hot corrosion attack were generally found to exhibit both oxidation and sulfidation. The hot corrosion morphology is typically characterized by a thick, porous layer of oxides with the underlying alloy matrix depleted in chromium, followed by internal chromium-rich sulfides. It is generally believed that the molten sodium sulfate deposit is required to initiate hot corrosion attack.

There are three categories of protective coatings used to protect gas turbine components: diffusion, overlay and thermal barrier coatings. Diffusion coatings are achieved by enriching the surface of an alloy with either aluminum, chromium or silicon. Overlay coatings are corrosion-resistant alloys that are designed for high temperature protection. Thermal barrier coatings are used to insulate the substrate from the heat of the gas

Figure B.5 Hot corrosion in GT 1st stage nozzle.

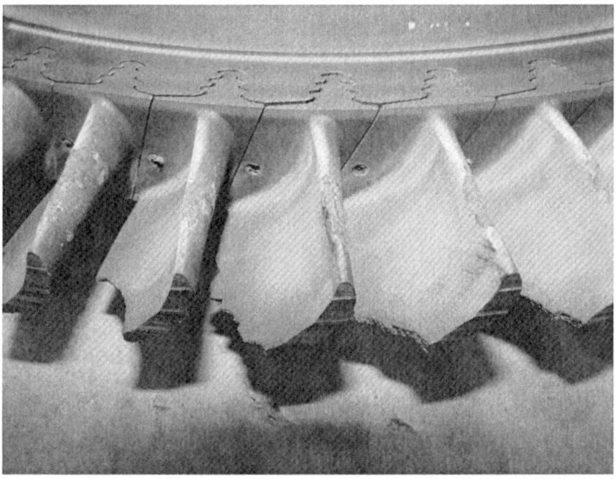

Figure B.6 Hot corrosion and blade tip rubbing.

flow. The coating system consists of an outer ceramic overlaid on an oxidation-resistant bond coat.

Common Hydrodynamic Bearing Failure Modes [5]

Babbitt fatigue: Fatigue damage may represent itself as intergranular or hairline cracks in the babbitt. The cracks may appear to open in the direction of rotation. Pieces of babbitt may spall out or appear to be pulled away in the direction of rotation. The cracks extend toward the babbitt bond line and may reveal the shoe backing.

Bond failure: Although not very common, bond failure can still be detected on some bearings. It is usually characterized by a complete separation of the babbitt from the backing material. High temperatures can cause the babbitt to melt near the bond line and then separate under the combined effect of dynamic loading.

Brinelling and fretting wear in bearing pads and shells: Bearing operating surfaces may suffer fretting damage while the shaft is at rest due to vibrations transmitted to the machine from external sources such as nearby machinery. Shipboard machinery is especially prone to this problem. If the machine is started up after fretting has occurred, the hard black deposit can cause wiping and obscure the original fretting damage. Similar damage can occur to machines in transit if the shaft is not clamped.

Babbitt wiping and overload conditions: A wiped bearing surface is where surface rubbing, melting and smearing is evident. This may be due to a variety of causes, including:

- Excessive load at start-up/run-down
- Insufficient lubricant
- Operational overload
- Loss of clearance
- Misalignment

Babbitt creep and thermal ratcheting: Thermal faceting affects only tin-based babbitt. Tin crystals are anisotropic, having different coefficients of thermal expansion in each crystal axis. Repeated thermal cycling results in faceting of the bearing surface due to a ratcheting process, which in severe cases may cause undulations in excess of 0.025 mm (0.001"). Rarely, cracks may form between the grains.

Cavitation damage: Cavitation erosion is an impact fatigue attack caused by the formation and collapse of vapor bubbles in the oil film under conditions of rapid pressure changes during the crank cycle in internal combustion engines. The harder the bearing material, the greater its resistance to cavitation erosion.

Electromagnetic current damage: In electrical machinery, damage to bearings from shaft currents can result from dissymmetry in the magnetic circuits. This dissymmetry could result from mechanical joints in the stator, a difference in the permeability of the magnetic material constituting two flux paths, or by difference in the effective length of these paths. To correct against such a problem, insulation is inserted between one or more of the bearing pedestals and the machine base. In addition, it would be necessary to insulate all piping, stairs, etc., which would otherwise complete the electrical circuit.

Electromagnetically generated currents are also known to occur in non-electrical machinery. The shaft or casing could become magnetized due to rub between rotating and stationary parts causing the generation of DC unipolar induced voltage. Magnetization of machine components could also be caused by welding on or near the machine, or through magnetic particle inspection. Damage due to electromagnetic type currents is generally more severe than that due to electrostatic type voltages. It often can cause welding of the journal pads to the retainer.

Scoring by foreign matter or dirt: Dirt in the oil system is one of the most common causes of premature bearing failures. The damage can range from light wear of the babbitt surface to severe scoring of the journal and bearing surfaces depending on many factors (see Figure B.7). Factors contributing to the severity of the damage include the size and amount of dirt, the bearing material, babbitt thickness, and bearing clearance as well as operating eccentricity. Dirt can get into the oil system during maintenance work or be ingested through breathers or air filters. Dirt can also come from metallic wear particles resulting from abrasive wear of moving parts or could be present in the oil distribution headers at the time of assembly and become dislodged during operation.

Corrosion damage: Corrosion in bearings can result from the chemical attack of the bearing alloy by corrosive compounds in the lubricant. These compounds can originate from external contaminants that find their way to the lube oil system, or from internal contaminants that form during operation of the machine as a result of lube oil degradation and oxidation. Corrosion due to oxidation of the lube oil is not as frequent today since lubricant additives have been developed to inhibit oxidation under severe operating conditions and for extended periods of time. Oxidation might, however, still occur if the oil analysis is not monitored and the oil is allowed to degrade. Oxidation may be accelerated by high temperatures and by exposure to air, water, or foreign materials in the oil. When lubricants become corrosive as a result of an accumulation of acids and peroxides, lead is removed from the copper lead bearing alloys. If both copper and lead are attacked and metallic sulfides are present in the corrosion products, this usually points to sulfur compounds as the corrosive element. The sulfur compounds can originate from the fuel system in internal combustion engines, seal leaks into the lubricating system in centrifugal compressors, or come as a result of an extended operation at high temperatures. These factors lead to the decomposition of sulfur compounds in the oil additives. Once decomposed, these compounds become very corrosive.

Figure B.7 Damaged jounal bearing.

Some bearing designs utilize an overlay over the copper alloy bearing to particularly guard against such a corrosive attack.

Pad flutter: Unloaded titling pads are also subject to flutter instability; a phenomenon often referred to as leading edge lockup or spragging. Leading edge lockup occurs where the leading edge is forced against the shaft and is maintained in that position by the frictional interaction of the shaft and the pad.

Porosity and blisters in babbitted bearings: Porosity issues typically fall into one of the following distinctive categories:

- Foreign inclusions. Cavities occur from inclusions such as slag or dross from the melting pot being trapped within the structure. This is the most common bearing defect. Such inclusions create a weak structure, sensitive to fatigue cracking.
- Absorbed gas. Cavities or blisters can also be caused by evolution of gases during the solidification of cast metal. When high lead babbitt is cast onto a cast iron backing, small cavities might develop. Photomicrographs can often reveal that the pinholes or cavities were caused by absorbed gas escaping from graphite flakes close to the babbitt iron interface. There is a higher probability of this occurring if the cast iron surface had not been degraphitized prior to casting the babbitt, thus, gases produced at graphitic foci during pouring were trapped within the babbitt microstructure.
- Dissolved gas. Some gases are soluble in molten metal and are rejected upon freezing, causing blisters. Molten

aluminum alloys have a marked tendency to absorb hydrogen from anything near the molten surface that can liberate hydrogen. This would include atmospheric water vapor or organic materials such as oil hydrocarbons. When the melt freezes, most of the dissolved hydrogen is rejected, but some becomes trapped in pores throughout the microstructure. As this trapped hydrogen can be expelled upon heating, engine operating temperatures are sometimes sufficient to raise the bubbles or blisters in an overlay structure.

- Powder impurities. Porosity can arise from impurities in the metallic powders which are compacted to produce sintered structure bearings. Each particle of organic matterlint, paper, wood splinters, etc., produces gases and carbonized residues at sintering temperatures, thereby creating a small void.
- Hydrogen electroplating. Gases, usually hydrogen, may accumulate under the electrodeposit, especially when the base support is steel. When certain metals, such as chromium copper, silver, and cadmium, are electroplated onto steel, the potentials are such that hydrogen tends to "plate out" concurrently with the primary metal. Some of this hydrogen gathers into bubbles in the plating solution; some diffuses into the steel base which, when unalloyed, is quite "transparent" to atomic hydrogen. When electroplating is completed, subsequent heating must frequently be employed to expel the hydrogen, which otherwise can lead to:

Wire-Wooling or Black Scab: If a large dirt particle (probably not less than 1 mm [0.04"] across) is carried into the clearance space by the lubricating oil and embeds in the bearing, a hard scab of material may form as a result of contact with the steel journal or thrust. collar. This scab will then cause severe damage to the mating steel surface, which is machined away and forms a so-called 'wire-wool' accumulation (see Figures B.8 & B.9). The susceptibility to scab formation depends on the nature of the lubricant and the composition of the steel of the rotor shaft or collar. Steels containing chromium or manganese in excess of 1% appear to be particularly susceptible to scab formation, especially in high-speed machines with bearing rubbing speeds over 20 m/s (4000 ft/min) and with oils containing extreme pressure additives.

Figure B.8 Journal bearing damaged by wire-wooling issue.

Figure B.9 Journal bearing, and journal damage caused by wire-wooling issue.

Rolling Element Bearing Failure Characteristics [6]

Rolling element bearings (see Figure B.10) are designed to have a long and useful life under design conditions. However, to maximize a bearing's operating life it must be properly installed, lubricated and maintained. If a bearing fails prematurely, it is important to determine the exact cause so appropriate modifications or adjustments can be made. The examination of a failed bearing can provide valuable clues to the cause of failure. The purpose of this guide is to assist failure analysts in properly identifying and analyzing 12 primary causes of rolling element bearing failures.

Figure B.10 Cutaway of a rolling element bearing in a housing.

1. Excessive Load: Premature spalling seen in the ball path.
2. Overheating: Symptoms of overheating are discoloration of the gold to blue. Temperatures in excess of 400°F can anneal the ring and ball materials. The resulting loss in hardness reduces the bearing capacity causing early failure. In extreme cases, balls and rings will deform. The resulting temperature rise can also degrade or destroy lubricant.
3. False brinelling: Elliptical path wear marks seen in an axial direction at each ball position with a bright finish and sharp demarcation, often surrounded by a ring of brown debris, is an indication of excessive external vibration. A small relative motion between balls and raceway occurs in non-rotating ball bearings that are subject to external vibration. When the bearing isn't turning, an oil film cannot be formed to prevent raceway wear. Wear debris oxidizes and accelerates the wear process.
4. True brinelling: Permanent sets of indentations seen in bearing raceways at rolling element pitch when the static load limit of the material is exceeded under load. The indentations reflect the rolling element shape and are surrounded by a build-up formed by displaced material.
5. Normal fatigue failure: Spalling or flaking of metal surfaces seen at the end of bearing life. This spalling can occur on the inner ring, outer ring, or balls. This type of failure is progressive and is normally accompanied by a marked increase in vibration.

6. Reverse loading: Balls show a grooved wear band. Angular contact bearings are designed to accept an axialy load in only one direction. When loaded in the opposite direction, the elliptical contact area on the outer ring is truncated by the low shoulder on that side of the outer ring. The result is an excessive stress and an increase in temperature, followed by increased vibration and early failure.
7. Contamination: Denting of found on the bearing raceways and balls.
8. Lubricant failure: Ball tracks and balls are discolored (blue/brown).
9. Corrosion: Chemical attack results in reddish/brown discoloration.
10. Misalignment: The raceway ball tracks are not parallel to the raceway edges. Abnormal temperature rises and premature wear of cage results from misalignment.
11. Loose fits: Effect: Circumferential wear and/or discoloration of the mounting surfaced are found. Cause: Loose fits can cause relative motion between mating parts. If the relative motion between mating parts is slight but continuous, fretting occurs.
12. Tight fits: A heavy ball wear path in the bottom of the raceway around the entire circumference of the inner ring and outer ring indicates a tight fit.

Tips for Analyzing Mechanical Seal Failures [7]

Performing a detailed seal failure analysis can be helpful in diagnosing, correcting, and preventing repetitive seal failures. The goal of the seal failure analysis is to identify a design flaw or operating condition that caused the failure so that it can be corrected. The first step in performing a failure analysis is finding a large clean table, in a well-lit room. Next, lay all the parts on a table with the seal drawing. Make sure to remove all the O-rings so that they can be carefully examined. Then examine all the parts carefully (Note: Figure B.11 shows what typical mechanical seal components look like in the new condition. By comparing the failed seal parts to their new respective parts, the analyst can readily identify the nature of the seal damage).

There are four fundamental questions that should be asked during the tear down and inspection$_7$:

Figure B.11 Mechanical seal components.

1. What does the seal damage look like? Is it chemical, mechanical, or thermal?
2. How does the damage affect the seal performance?
3. What does the seal damage indicate about the seal's design and operation conditions?
4. What corrective steps should be taken to prevent the observed damage from recurring?

Common Seal Failure Modes

Here are the failure modes commonly found during mechanical seal shop inspections:

Overall Chemical Attack, which leaves parts appearing dull, pitted, eroded, or falling apart. Corrosion of seal components results from using the wrong materials of construction or failing to use a barrier fluid to isolate the seal from the corrosive fluid.

Remedial action can include: 1) obtaining a complete chemical analysis of the liquid being sealed and then upgrading the materials of construction to meet the chemical requirements, or 2) using double seals or flushing a single seal with an external source with a clean fluid that's compatible with the process to isolate the seal from a potentially corrosive environment.

Fretting Corrosion, which causes leakage at the secondary seals and damages the shaft sleeve directly under the secondary seal. The site damaged due to fretting, which appears pitted and shiny bright, results from the repeated back-and-forth motion, due to assembly issues and/or vibration, of the secondary seal over the shaft sleeve.

Causes of the excessive vibration of secondary seals are eliminated by making sure shaft runout, deflection, and end play are within recommended limits. Protective hard-faced alloys or coatings can be applied at the location of secondary seals. If possible, TFE, V-rings, wedge rings can be replaced with more forgiving O-rings. Another design option is replacing pusher-type seals with non-pusher type seals that have static secondary seals.

Damaged elastomers, which appear swollen, hardened, bubbled, broken, or blistered, result from the chemical attack or excessive temperatures. The most likely causes are incorrect material selection or barrier fluid contamination or loss.

The fluid being pumped should be chemically analyzed and the elastomer material selection reevaluated. Frequently, the presence of trace contaminants, originally overlooked when specifying the seal, is the cause of the problem. If a suitable elastomeric material cannot be found, then the seal must be isolated using an external seal fluid or replaced with a seal design using a static secondary seal.

O-rings that appear cut or peeled may have been extruded. The cause of this type of damage may be due to excessive pressure or temperature, chemical softening, or improper shaft and groove sizing. The cure can be 1) a reduction in pressure or temperature, 2) change in the seal design, O-ring material or durometer, or 3) proper sizing of the shaft or groove on the machinery.

Leaching, which gives tungsten carbide and ceramic faces a dull and matted appearance, can reduce surface hardness by 5 points or more on the Rockwell A scale. To correct the problem, upgrade the seal base material. Another way to eliminate the problem is to provide a seal arrangement that provides a buffer fluid at the seal faces, such as a single seal with an external flush or a double seal with a suitable barrier fluid.

Face distortion, which results in a non-uniform face wear pattern, frequently causes excessive leakage. Improper assembly or uneven gland nut torquing can result in nonuniform face loading. Other causes include: 1) improper cooling induce thermal stresses and distortions, 2) poor surface finish at the face of the seal housing, due to corrosion or mechanical damage, and 3) driver to pump misalignment.

Seal faces could be re-lapped to remove face distortions. Other corrective actions include using flexibly mounted stationary seal faces to compensate for gland distortion, checking the driver to pump alignment, and retorqueing the gland to recommended values.

Seal face deflection, which appears as an uneven or narrow wear patter, can be either concaved or convex. Seals with either condition are not stable under cyclic pressure conditions. Causes may be improper stationary seal face support, swelling of secondary seals, excessive pressure, inadequate balancing of hydraulic and mechanical loads on the primary seal faces.

Corrective and preventive measures include checking the seal design's operating limits and consulting the seal manufacturer about other seal designs., considering a flexible mount for the stationary seal face and replacing carbon seal faces with seal faces having a high modulus of elasticity, such as silicon carbide.

Tracking, which is a wear pattern on the larger face that is wider than the thinner face, caused by misalignment, pipe strain, cavitation, poor pump condition, or excessive pressure. The problem is corrected by checking and adjusting the seal in its housing and ensuring operating and machine conditions are correct. In instances where seal face alignment is not possible, a design with a static secondary seal may provide enough compensation.

Heat checking is the presence of fine-to-large cracks that radiate outward from the center of the seal ring. These cracks act as a series of cutting and scraping edges against the opposite seal face. Heat causes this condition due to a lack of cooling, vaporization at the seal faces, or excessive pressures or face velocities.

Temperatures can be reduced by flushing or cooling the seal faces and eliminating overloading.

Blistering, a symptom characterized by small circular sections that appear raised on carbon seal faces. Blisters separate the seal faces during operation and cause high leakage rates. Blistering commonly occurs in seals in high viscosity applications that are started and stopped frequently.

Possible actions in high viscosity applications are either increasing the sealing temperature or substituting a less viscous buffer fluid. Other actions include eliminating frequent starts and stops or substituting a non-porous seal face material, such as tungsten carbide.

Appendix C

Common Machinery Failure Modes [8]

Regardless of the lengths we go to purchase reliable designs, install, monitor, and repair them properly, and maintain proven procedures, machines will still fail in a variety of ways. Experience has shown us that certain failure modes occur more than others. For example, bearing and seal failures are probably the most common failure modes you will encounter, while impeller and shaft failures are rarely encountered.

Failure modes can be considered either primary or secondary in nature. A primary failure mode is an initial internal condition that can lead to subsequent catastrophic failure. For example, a plugged suction strainer can be considered a primary failure mode. By itself, a plugged strainer is not catastrophic, but it can eventually lead to cavitation (secondary effect) and eventually a bearing failure (secondary failure) due to low flow conditions.

Some other common machine failure modes include:

- Pluggage (usually a primary failure mode)
- Erosive Wear (usually a primary failure mode)
- Fatigue (primary or secondary failure mode)
- Bearing failure (primary or secondary failure mode)
- Rubbing (primary or secondary failure mode)

By learning to recognize these common failure modes in the field, we can become better failure analysts. Once the failure mode is identified, we can then attempt to either prevent their recurrence or minimize the extent of secondary damage they may cause if gone undetected. Here we will discuss these failure modes in some detail and examine actual field photos of each type of failure mechanism.

Figure C.1 Plugged suction screen on compressor inlet piping.

Figure C.2 Plugged flue nozzles from a gas turbine.

Pluggage

Pluggage occurs due to the gradual build-up of solids in the flow path of the process piping or machine component. Symptoms of pluggage include the loss of flow or pressure and can sometimes lead to flow related vibration. Figures C.1 & C.2 show two representative photos of pluggage that resulted in failures.

Figure C.3 Impeller worn out due to erosion.

Figure C.4 Expander wheel eroded due to liquids in the inlet gas.

Erosive Wear

Erosive wear occurs whenever an erosive fluid removes material from the component whether it be a piece of piping or a pump. Erosive fluid typically results from an abrasive component within the process fluid that causes removal of material much like sandpaper. Symptoms can include vibrations due to imbalance, flow related vibrations, loss of flow or pressure, or the loss of overall efficiency. Figures C.3 and C.4 show two photos of component damage resulting from erosive wear.

Fatigue

Fatigue occurs whenever a material is subjected to cyclical loading and unloading for an extended period of time and the resulting stress levels are above some limiting design threshold. In other words, when the dynamic stress loading exceeds a certain threshold stress level, microscopic cracks will begin to form at stress risers, or concentrators and then grow over time. Fatigue can happen over time, or very rapidly. Fatigue failure that occurs over many cycles usually has two distinct stages. The first stage is a propagation of the crack over a number of cycles and is evident by beach marks. Eventually, the crack reaches a critical size, and the component fails rapidly. Causal factors such as high vibration and pressure pulsation levels, and unsteady flows are often to blame. Recall, a high static load with a cyclic load on top can result in fatigue failure too. Figures C.5, C.6, and C.7 show some component fatigue failures.

Figure C.5 Pipe clamp bolt fatigued due to high shaking forces.

Figure C.6 Engine connecting rod bolt fatigued due to faulty bolt design.

Compressor Blade Fatigue Example

The leading edge of a centrifugal compressor blade failed (Figure C.8) due to fatigue, which led to a rapid increase in high 1x vibration as a result of the imbalance of the impeller suddenly increasing. The manufacturer concluded that the combination of a thin casting thickness at the leading edge of the blade and vane pass pulsations led to a resonant condition at the blade. A resonant condition requires that the components natural frequency (dictated by geometry, material property, and mass all design issues) coincided with an excitation (pulsation due to vane pass).

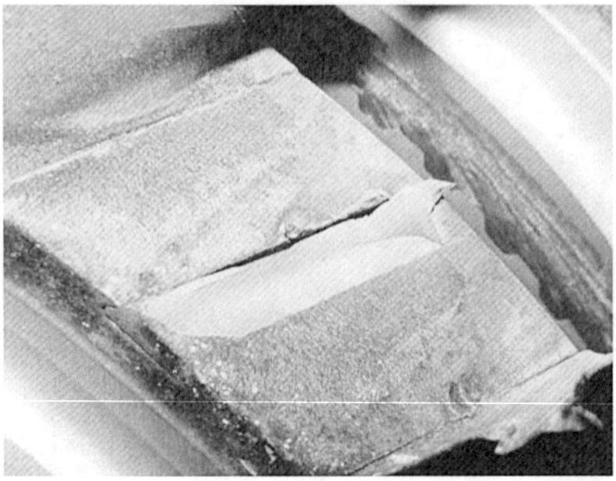

Figure C.7 Axial compressor blade fatigue due to blade flutter.

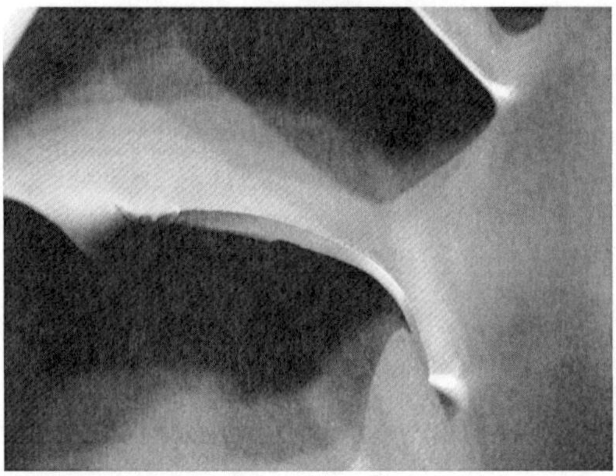

Figure C.8 Alternate angle of failed blade.

Hydrodynamic Bearing Failure Examples

Hydrodynamic bearings are highly engineered components that require proper design, proper installation, and proper maintenance for successful operation. Although these types of bearings are simple in design and construction, they can fail for a variety of reasons. Failure could arise from poor lubrication, overloading, misapplication, and fabrication defects. Normally, bearings display symptoms of distress before they fail. Symptoms of a failing bearing include higher vibration levels and rising bearing or oil sump temperatures. However, once a bearing begins to exhibit failure symptoms in practice it is difficult to predict when it will actually fail completely. Figures C.9, C.10, and C.11 show a few representative examples of bearing failures.

Figure C.9 A main bearing from reciprocating engine is beginning to fail after the grout failed.

Figure C.10 Tapered land thrust bearing beginning to fail due to compressor surge events.

Rubbing

Rubbing occurs whenever a rotating component contacts a stationary component inside a machine. Rubbing is usually a secondary failure mode resulting from the loss of clearance between internal components after a bearing failure. Excessive shaft deflection from vibrations or thermal growth can lead to rubbing too. Symptoms related to rubbing normally include high vibration levels and noise. Figures C.12 and C.13 show two examples of rubbing failures.

Figure C.11 Wire-wooling bearing failure due to high chrome shaft and oil contaminants.

Figure C.12 Expander wheel on the right shows sign of rubbing due to excessive axial movement. (New wheel is on the left.)

Unique Failure Modes

Every piece of equipment has its own unique design issues and challenges in operation, and each has its own unique failure modes. So, it is recommended that you talk to the original equipment manufacturer about known design issues, operating problems, and failure modes before their machines are installed in the field.

New designs typically present greater challenges to machinery and operating professionals as they may be unpredictable in their response to

Figure C.13 Axial air compressor blade fatigues due to tip rubbing.

process or environment changes. If machines are a completely new design, designers may have to rely on computer modeling tools (CFD, FEA, etc.) to determine the critical locations where stress, temperature, heat transfer, etc., may reach recommended limits. This is a challenge as computational models are actually quite restrictive in accounting for all variances that might be encountered. It may later be determined that design assumptions were incorrect or that process conditions were wrongly predicted.

If a new design fails to perform reliably you should work with the original equipment manufacturer to determine the nature of the problem. This may require gathering field data and performing failure analyses on failed parts in order to identify the root cause of the problem. The ultimate problem resolution may require a combination of field data gathering and failure analyses, field testing, and analytical modeling.

References

1. Perez, R. X., Using Fault Trees to Determine the Root Cause of Rotating Equipment Failures, *Proceedings of the Tenth Pump Users Symposium*, 1993, pp. 184-195.
2. *Cause Mapping®: Effective Root Cause Analysis*, Technical Brief, Think Reliability.com.
3. Source: FORD, Team Oriented Problem Solving
4. Lai, G. Y., *High-Temperature Corrosion and Materials Applications*, Chapter 9, Hot Corrosion in Gas Turbines, ASM International, 2007

5. Zeidan, F. Y. and Herbage, B. S., Fluid Film Bearing Fundamentals and Failure Analysis, pp. 161-186, *Proceedings of the Twentieth Turbomachinery Symposium, Sept. 1991.*
6. Barden Precision Bearings, Bearing Failures: Causes and Cures, www.schaeffler.com,
7. Foszcz, Joseph, L., Senior Editor, *Preventing Mechanical Seal Failures*, Plant Engineering Magazine, Sept. 1992, pp. 47-51.
8. Perez, R. X., *Troubleshooting Rotating Machinery,* Chapter 3: Machinery Design Issues and Failure Modes, pp. 37-52.

17

Investigation and Resolution of Repetitive Fractionator Bottom Pump Failures

By Abdulrahman Alkhowaiter

Figure 17.1 Fractionator bottom pump.

Introduction

Since their commissioning in January 2015, the Refinery 112-G-0019 A, B Fractionator Bottoms Pumps (see Figure 17.1) had experienced several seizing-related failures during startups. These are Model 10HED28DS (see Figure 17.2) high-temperature service pumps handling service fluid of hot oil rated at temperatures of 300 to maximum 355 DEG.C. Both pumps had several failures in 2015 due to this common factor and were sent to the OEM shop for repair.

Robert X. Perez (ed.) Maintenance, Reliability and Troubleshooting in Rotating Machinery, (333–342)

The March 2015 Root Cause Failure Analysis report covered several important factors on the initial failure causes of these pumps. Report findings were:

1. Improper hot pump warm-up during startup. This caused uneven thermal heating of pump casing leading to internal misalignment resulting from different temperatures of casing & components.
2. Insufficient size orifice [7 mm] on warm-up lines, which delays heating of pump.
3. Mechanical Seal Plan 32 is adding cold ambient temperature fluid to pump; this piping should be isolated during startup warming procedures.
4. Refinery operating procedure did not monitor pump casing Delta-T per manufacturer's requirement of monitoring Delta-T of top/bottom pump casing temperatures before startup.

List of Additional Failure Inherent Causes to Be Rectified

In addition to the above findings, the investigation team highlighted some additional Root Cause failure modes that they thought were contributing to the hot service pumps failures. These are issues that could not be ruled out without additional inspections. Therefore, it was recommended that the following issues be addressed:

Internal cold misalignment of rotor to stationary parts can be caused by many factors. Standard manufacturing procedures are not accurate. This leads to rotating parts that are offset internally and very close to rubbing with stationary elements (see Figure 17.2). The vibration spectrum recorded on Oct. 9, 2015 (Figure 17.3), seem to support the idea that internal rubbing was occurring. Once some minor temperature difference occurs, the shaft immediately seizes in casing. This includes incorrect bearing housing to casing accuracy. Also, internal misalignment or excess clearance of stationary parts, such as covers, to the casing. These must be checked by destacking pump at shop and taking the measurements listed below. All alignments are recorded, and the maximum allowable offset was found to be only 0.002 inches. Max allowable face runout of casing flange contact surface with covers is 0.002 inch. Same for the covers flange faces.

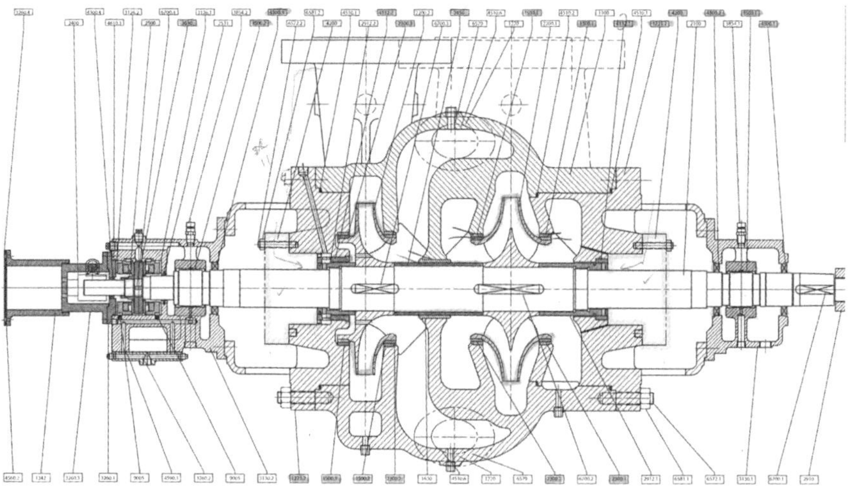

Figure 17.2 Pump X-section showing rub locations.

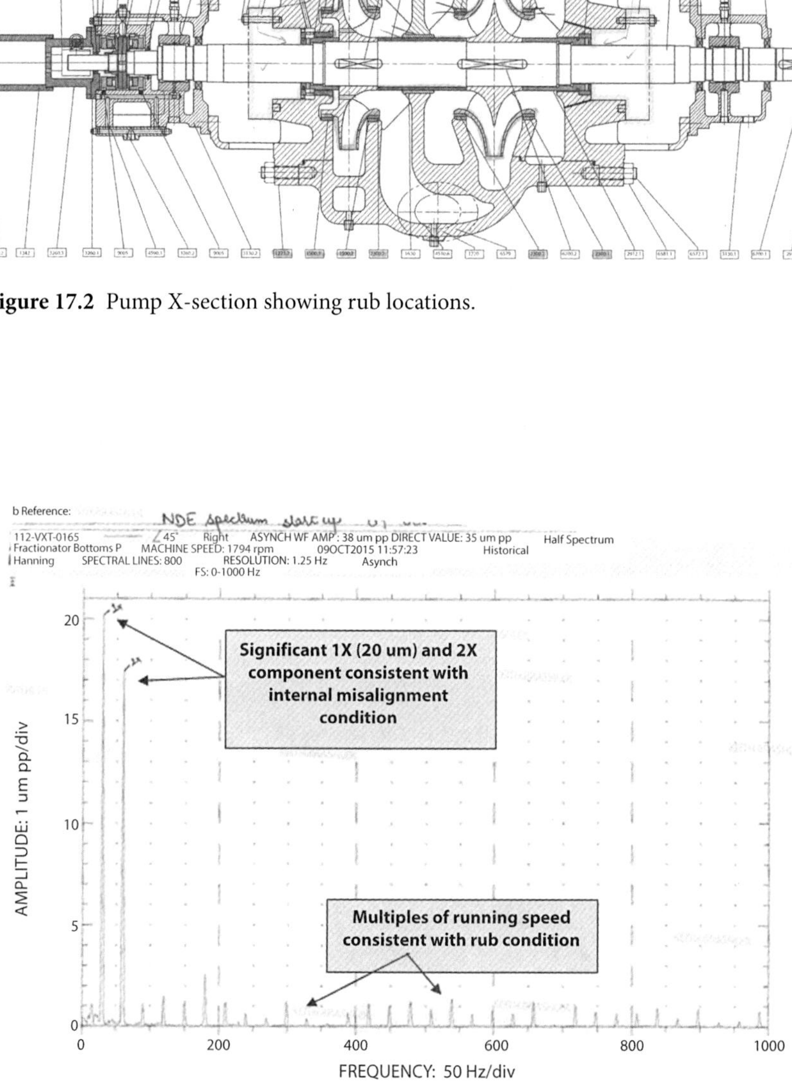

Figure 17.3 Frequency analysis of bearing X-Y probes vibration before mechanical upgrades (Oct-9, 2015).

Register diametrical fit clearance of covers to casing, is max allowable of 0.004 inch.

Key Shop and Field Pump Measurements

1. **Gravity sag deflection of rotor midspan at high temperature service.** This makes impellers operate non-concentric to stationary wear rings, making it easy to rub especially at the center bushing. Preferred to offset the rotor-bearing concentricity to pump casing at shop, by 0.005 inch vertically upward to compensate for hot sag in middle of rotor for this size pump.
2. **Insufficient clearance on rotating and stationary components.** For example, the outer diameter of stationary wear rings fitting to casing internal diameter must have adequate clearance (Figure 17.4 and 17.5). Otherwise, pinching occurs at operating temperature and loss of design clearances with rotating impeller. Need minimum 0.007 inch diametral clearance. Rotor Balance drum [piston] concentricity and correct clearances, to casing when assembled is also critical. **Impeller Wear Ring:** Insufficient clearance on impeller wear ring to casing wear ring is another possibility especially if thermal growth differences lead to loss of dimensions when

Figure 17.4 Impeller wear ring clearances larges enough to prevent rubs during startups and upsets.

Figure 17.5 Casing stationary wear rings can rub on rotating impeller.

hot. Recommended API-610 paragraph 5.7 clearance for 24.0 inch diameter impeller is 0.036 inch diametral clearance. **Center Bushing** clearance should be in range of 0.012 to 0.015 inch diametral.

3. **Verify Close Clearance Wearing Parts Hardness to Reduce Galling:** As per API 610 the minimum required hardness difference shall be 50 HBN. For materials with hardness higher than 400 HBN, no need for any difference.
4. **Distortion due to pump baseplate flatness or unequal height feet:** Pump supports on baseplate may not be flat within 0.25 millimeter maximum. Also, one support may be higher than another. When pump is bolted down, this then causes a soft foot condition that twists pump casing, leading to internal misalignment and rubs. Check this by performing a laser sweep of the pump base footing to ensure equal height of its footing and flatness. A field machining contractor can be hired to perform field laser check and machining of base.
5. **Piping strain on pump discharge and suction flanges.** This force distorts pump casing, leading to internal rubbing. This has already been found and rectified by maintenance. But still some remains to be fixed.
6. **High coupling misalignment between pump and driver.** After all the above repairs are performed on pump and baseplate, the final step is to perform a hot alignment laser check of the coupled machines (see Figure 17.6). This is done by

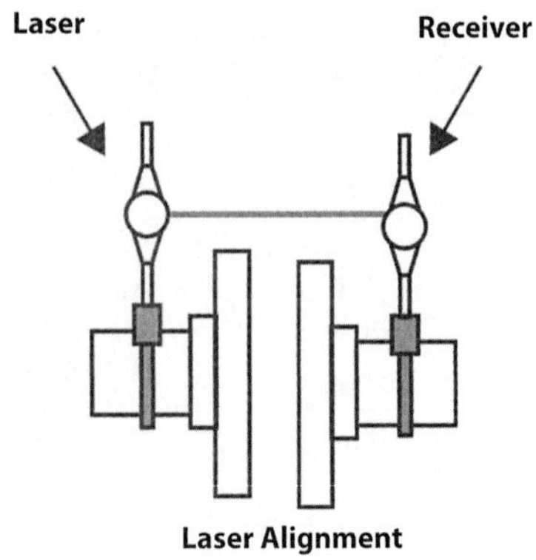

Figure 17.6 A laser alignment tool was used to check the driver to driven machine alignment under operating conditions.

operating the pump at normal condition for 12 hours or more, then shutting down immediately and applying laser alignment tool on coupling behind hubs, without allowing cool down of pump. The coupling spool is not removed, as laser tools do not need its removal. The new desired cold figures are now applied, and original alignment readings are deleted.

7. **Human Factors in Machinery Startup:** It is important for plant designers and rotating equipment engineers to understand that writing long, complicated machinery startup procedures for operations personnel **will lead to accidents and safety deficiencies**, because the longer the list of actions and check points, the more human errors or neglect occur. It is imperative for the reliability engineers to perform a study on pump bad actor equipment to reduce the startup procedures by making design changes to pump and associated auxiliaries, piping, addition of reliable automatic devices, and developing innovative methods which achieve same startup reliability and safety.
8. **Add Orifice 6 mm to the top entry warm-up line of both pumps:** When excess warmup flow enters this top line, it reduces the bottom entry flow, thus leading to bypassing of correct circulation direction. This leads to incorrect heating of standby pump casing. Add 6 mm orifice in the line at location shown in the drawing below.

9. **Novel Hydrocracker Pump Startup Procedure Eliminates Rub:** The Hydrocracker operation superintendent in August 2015 implemented a successful startup with an innovative method on the turbine driven pump. This method involved filling the G-19 pump casing with diesel fuel, then starting the steam turbine at 500 rpm while monitoring pump temperature and vibration. After 30 minutes the pump warm-up line valves were opened allowing hot oil to enter pump (see Figure 17.7). Next, the main suction valve is opened, and pump driver speed is ramped up to 1500 rpm for 15 minutes, and if pump readings are smooth and casing outer temperature reaches at least 50% of pump rated liquid temperature, then full speed at 3575 rpm is applied. This is an acceptable procedure. However please note, the following safety guidelines: This method should only be used if pump rotor is found to rub with internals during normal hot startup. **Warning: Overpressure of casing & piping:** When opening the warm-up lines, these will pressurize the casing and the suction piping up to the closed suction valve until reaching full discharge pressure, unless the suction main inlet valve is cracked open, or if its bypass line is cracked open. In this case it's very critical to maintain a partial opening of the

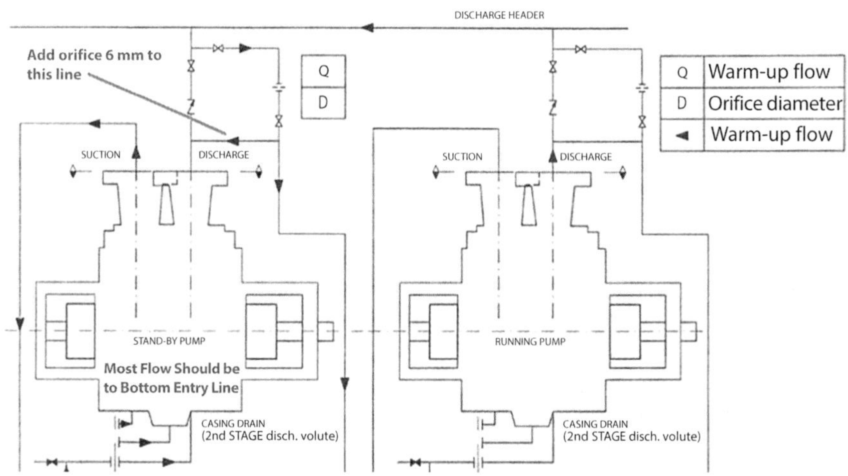

Figure 17.7 Warm-up piping modification to top entry line.

suction valve or its bypass line so as not to reach full discharge pressure in casing and suction line.

Note: With any blocked-in pump, it is important to recognize that such pumps cannot be left operating in a closed check valve position for long operating periods of time because a shaft rotating in a pump transfers friction heat that will continuously heat up pump and casing and can be dangerous. Therefore, senior engineers should be at site during such starts. In addition, a second possibility is to utilize a barrel of lube oil ISO-VG-32 light oil instead of diesel fuel. This has more seizure resistance and is a safer fluid. While this method is suitable for a variable speed drive such as steam turbines, it will be more sensitive to start up errors in case of motor driven constant speed pumps due to fast startup. For safety purposes, it's important to fill pump casing without leaving any air space as pump must be carefully primed 100%. Air pockets should not be trapped in such service pumps.

Conclusion [1]

Hot service pumps are well known for having reliability problems, particularly seizure failures, which includes both hot hydrocarbons and boiler feed water pumps. Implementing all the above checks and repairs was needed to eliminate future possibility of internal seizing of such pumps, as all inherent defects will have been rectified.[1] The fact that with both G-19 A, B pumps down, 75% of refinery output is lost means that the efforts and expense to raise these pumps' reliability will have immediate payback. It is strongly recommended to perform above repair checks the next time these units experience a seizure failure or require shop maintenance.

Actual Findings

The pumps were modified and repaired per the recommendations in this chapter and since these improvements, zero pump seizures have occurred from Nov. 2015 to 2021. Out of eight possible root causes for seizure, the following five were actually identified as active failure causes and fixed at the OEM pump shop and at site by Refinery piping specialists. It should be noted that OEM was removing blame from themselves by considering all past G-0019 A, B failures on “Piping Strain” as this absolved them of responsibility for the six past failures.

The actual findings discovered during the repairs were all addressed, which led to a 100% improvement in pump reliability after November 2015. The following improvements were aimed at resolving all inherent G-0019 A, B failure causes:

1. Internal pump casing stationary parts alignment to rotor improvement by remanufacturing at Pump shop all components for correct geometric concentricity and face runout. Root Cause of this Failure mode was poor manufacturing accuracy and tolerances in the original pumps supplied by manufacturer.
2. New warm-up improvement with orifice to balance flow. The warm-up design has been improved by following the March 2015 report on warm-up changes, and the September 2015 report requesting additional warm-up orifice modifications. Root Cause of this failure mode was faulty warm-up piping design by pump OEM.
3. Elimination of pump piping strain by cutting inlet pipe and aligning. Piping strain was considered medium, not an extreme piping strain. Root cause of this strain was Lump-Sum-Turnkey (LSTK) contractor piping fabrication errors.
4. Additional internal clearance for close clearance parts. The consultant recommended additional operating clearance on the pump internals to meet API limits for hot pump service. This was applied at pump repair shop.
5. Incorrect Pump Metallurgy selection in Data sheet: soft casing wear ring material was used against soft impeller wear rings, as selected incorrectly by the LSTK designer and he did not follow correct selection table of Refinery pump standards. In addition, the pump manufacturer did not provide correct combination of hard vs. hard or hard vs. soft in the stationary-rotating part. Using soft vs. soft parts leads to seizures because they will gall upon contact. Root cause is incorrect design selection by LSTK contractor, lack of proper pump data sheets review during design, and errors by OEM in not providing correct material hardness combination in a hot service pump.

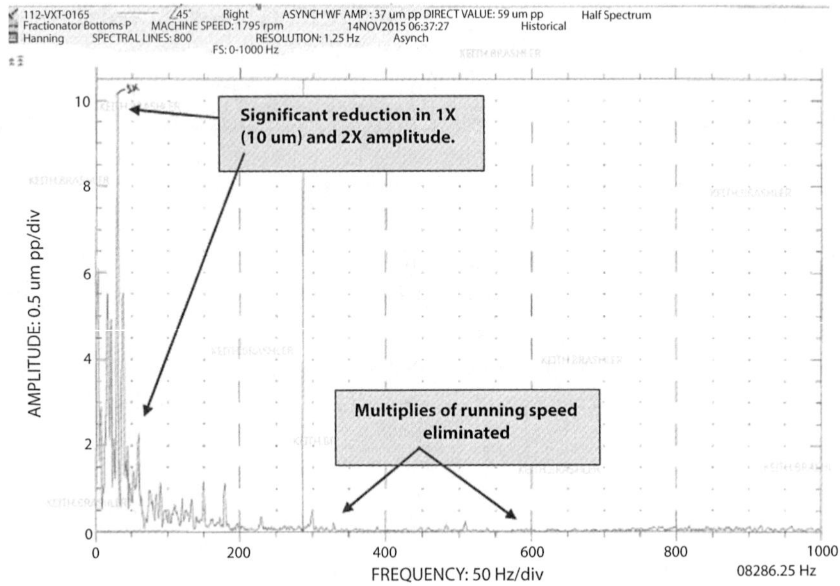

Figure 17.8 Frequency analysis: of non-drive bearing using X-Y probes (Nov-14-201).

Effect of Improvements on Pump Radial Shaft Vibration

Figure 17.8 shows the NDE (Non-drive End) proximity probe vibration spectrum taken after all improvements were applied. The maximum vibration level after all reliability improvements were implemented dropped to 1.50 mils overall, indicating a 50% reduction in vibration at normal flows and pressures. Current pump field data show that vibration levels have hardly changed since the initial November 2015 startup.

Reference

1. For further reading, see Turbomachinery article "Maintenance & Troubleshooting of Single Stage pumps" by Ed Nelson.

18

Reliability Improvements Made to 6000 KW Water Injection Pumps Experiencing Wear Ring Failures

By Abdulrahman Alkhowaiter

Summary

Impeller wear ring failures during service on five identical pumps led to high vibration, seizures, and reduced output. The pumps have 1800 RPM motor drive, through gearbox to 4960 RPM, with 6000 KW output shaft (See Figure 18.1). The water is 15,000 TDS water, but has suspended solids. Major internal failures occurred on average from 9 months to one year of operation since commissioning in 2005. The design solutions included a change in Wear Ring fit to impeller method from OEM original specifications and increase in surface coating hardness at wear rings. Result was increased MTBO from 1.0 year to 4 years MTBO, which is a 400% increase.

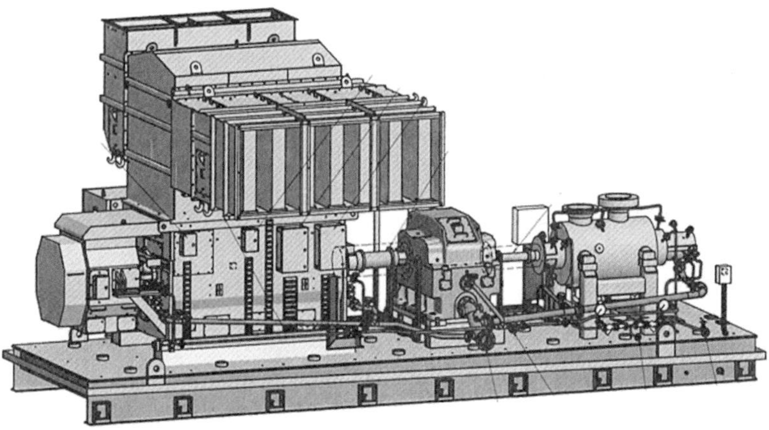

Figure 18.1 Water Injection Pump Drive Train: An electric motor (far left) drives a gearbox, which drives a centrifugal pump (far right).

Robert X. Perez (ed.) Maintenance, Reliability and Troubleshooting in Rotating Machinery, (343–352) © 2022 Scrivener Publishing LLC

Sequence of Events

On January 8, 2008, Operations asked for the reliability group's support in investigating the low discharge flow problem on the G-129 pump. They explained that the discharge flow maintained a stable value of 3700 USGPM till Sat. 5th and then it dropped to 3200 USGPM till it tripped on high vibration at the outboard bearing on Mon. 7th. This indicated that there is a real low discharge flow problem that may be related to internal mechanical problem with the pump or the system.

On January 9, 2008, the problem recurred. Again, there was a huge reduction in flow rate of the pump from 3700 to 3200 US GPM, which is an indication of internal leakage between the stages as a result of losing the wear ring again. This problem was encountered before, and the pump was sent to you for inspection and repair. Further to the below e-mail please note that the power load consumption of the subject pump reduced from 6.1 MW to 5.5 MW. When opened in the shop, the pump internal parts were exposed for inspection (see Figures 18.2 and 18.3) and the following was observed:

- Fine sand particles suspended in water were noticed clearly inside the casing.
- Signs of wear were found on some of the internals parts like the vanes and the impeller eyes.
- Wear of balance drum and the wearing rings. Severe corrosion of the casing head studs.
- All wear rings had cracks at Butt weld; one had snapped off and entered other parts of pump.

Figure 18.2 Pump internals.

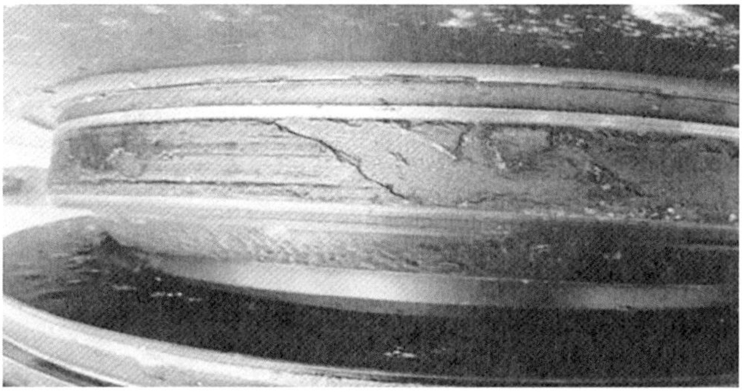

Figure 18.3 Close-up of pump wear rings.

On January 19, 2008, the G-129 Pump was disassembled for inspection in the shop. The following observations were made: the fourth stage impeller lost its rotating wear ring, and fragments were found in the pump casing. The last shop overhaul wear ring fit method is a shrink fit plus spot welding at sides of ring to impeller. By careful investigation of the failed wear ring, a 75% length corroded crack was found emanating from the weld zone on one fragment. **This is proof that the root cause is due to welding zone crack leading to wear ring failure in-service.**

The manufacturer original design used grub screws to hold wear rings against axial forces. The grub screws failed in service, prompting the shop to substitute a new design, which consisted of flash butt welds at impeller wear ring faces after shrink fits were applied. The manufacturer's standard of axially installed set screws was deemed to be inferior through shop experience. Unfortunately, both set screws and welds can cause stress cracks, since these components are shrink fitted with a residual circumferential hoop stress. The impeller-to-wear ring holding procedures must eliminate the stress concentration, we developed a new Idea:

- The Impeller Wear Ring land OD is machined with a 0.010 inch diametral taper (Figure 18.4). The wear ring ID is also machined with a tapered bore, 0.010 inch total taper. A diametral shrink fit of 0.0005 inch/inch wear ring diameter is used to hold the wear ring to the impeller. This design captures the impeller wear ring axially without adding a stress concentration. Torque transmission is through the shrink fit.

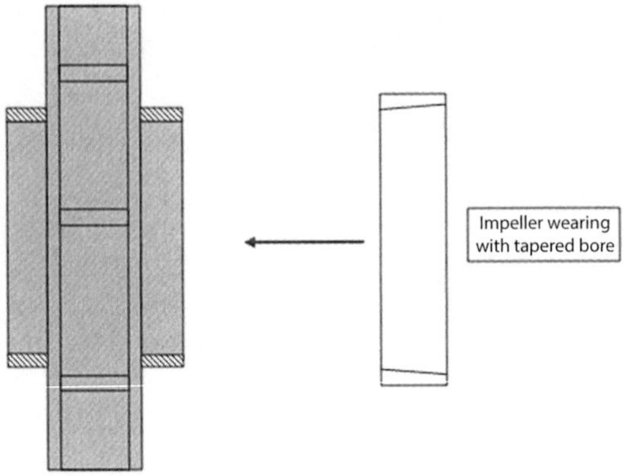

Figure 18.4 Wear ring geometry with tapered bore.

- There is a small female taper machined into the pump impeller to mate exactly with the wear ring taper. The taper is such that at one point, the height is 0.010 inch greater than other side of wear ring. This axial interference makes it practically impossible for wear ring to slide off impeller axially.
- Next: An additional shrink fit is added to hold impeller to counter differential thermal growth during service, and to counter centrifugal stress relaxation. Use 0.0005 per inch diameter. To remove the wear ring [or install it] then it must be heated so that total diametral growth is enough to slip over the impeller.

New Design Proposal of Eliminating Grub Screws or Flash Butt Welding

Example: Wear ring ID = 8.0 inches. Apply Taper Fit Principle

Taper Fit step height: 0.010 inch.

Diametral Shrink fit additional= 0.0005 per inch ID = 0.0005 x 8 = 0.004 inch.

Total thermal growth required to fit wear ring to impeller: 0.010 step + 0.004 radial fit + 0.003 [clearance to allow fitting] = 0.017 inch

Total thermal heating growth = 0.017 inch.

For SS-316 or Nitronic wear rings thermal growth coefficient is 9×10^{-6} inch/inch Deg F

Delta S = 0.017 = [T-final-Shop Ambient 75 Deg F] x 8.0 inch diameter x 9×10^{-6}

Delta -T required = 0.017/ [$8 \times 9 \times 10^{-6}$] = 236 Deg F

Final temperature required of wear ring is: Ambient + Delta= 75 + 236 = 311 Deg F.

Add 30 Deg F as safety margin= 340 F. Use laser gun to ensure correct temperature.

Upgrade Options

We discussed another design solution for improving the pump mechanical reliability, and the pump consultant proposed alternative design of wear ring attachment to impeller:

1. The wear ring is welded fully on circumference [both sides] directly to the impeller with full circumferential welds of ¼ x ¼ inch width.
2. However, note that there is no shrink fit imposed; this eliminates high stress as cause of failure. The wear ring thickness is also increased by 1/16 to 3/32 inch to increase strength.
3. After welding to the impeller, the wear ring is hard faced on the OD by applying welded hard face of Stellite.
4. A weld stress relieving procedure is followed after welding to reduce thermal stresses.
5. Final Machining/Grinding of the assembly is done to achieve final dimensions.

The two options of either 1) directly welding the wear rings to impellers or 2) using a tapered wear ring fits were presented to the end-user and senior pump engineer for their review. It was decided that the fully welded wear ring design would be the best option since the wear rings were already stocked and had been proven to provide relatively trouble-free service. It was thought that, with the welded wear ring option, the wear rings would become an integral part of the rotor, meaning there would be a near zero probability of any component breaking off the pump rotor in service.

Detailed Analysis of Problem & Solution Related to All Pump Wear Rings

Details: The 6000 KW injection pumps were installed in 2004 at water injection pumping service. The water suction temperature is about 190 DEGF, at 300 psig suction pressure from wellhead. Pump RPM is 4964. This 2900 PSIG discharge pressure pump had numerous failures MTBF of 1.0 years average for three units due to two reasons. First, incorrect data sheet data on pumped formation water only about 15,000 TDS (total dissolved solids) led to sand erosion of pump internals. The second reason is that the impeller wear rings would fracture during operation due to stress corrosion cracking at stress raiser of axial locking screws on impeller wear rings.

The successful internal design solution for wear rings stress corrosion failures was to use Nitronic-60 wear rings with low shrink fit with total diametral shrink of only 0.003 to 0.004 inch maximum. Following details of repair were supplied by our repair specialist:

1. The wearing rings were first fully welded [back and front of wear ring] to the CF3M material impellers landing using Inconel 625 welding rods, and during this time, a dummy stub shaft is shrink fitted temporarily into impeller bore to eliminate shrinkage of bore from welding.
2. At this step, some light machining on wear ring OD is done.
3. Then the impeller wear rings OD were coated with Tungsten Carbide using HVOF Metco spray powder #2002 or #5803. The Tungsten Carbide spray coating thickness, after final grinding was in the range of 0.015 to 0.020 inch maximum.
4. Wear ring OD has a 0.075 inch radius on edges to reduce stress raiser, and Tungsten Carbide coating is kept 0.062 inch away from wear ring edges to reduce stress riser on coating layer. Wear ring ID has 40 Degree chamfers for welding, about 3.5 mm x 3.5 mm.

This completely eliminated the wear ring infant mortality failures since 2008; no wear ring failures occurred again on three modified identical pumps until failure in late 2012. Note that a single wear ring failure in the past caused immediate internal damage and high vibration requiring overhaul of all sister pumps overhauled at our shop. The alternative design concept solution was to use a tapered inner bore for the impeller wear rings and opposite taper OD on Impeller Lands, with no welding used. The taper bore idea was to be used with a conventional shrink fit used on impellers. This eliminates the need for axial fixation screws or spot welding to hold impellers. These both cause fractures in service.

The reason that impeller wear rings in general require axial fixation in addition to normal shrink fit is due to:

1. The impeller wear ring may have a differential thermal expansion of bore during operation if its material is different than that of the impeller.
2. The impeller wear ring typically is exposed to higher centrifugal stresses at impeller land [partial loss of shrink fit] than the impeller due to its large diameter and due to stiffer construction of impeller which has less elastic deflection compared to ring.
3. The impeller's centrifugal force deflection also causes an axial force trying to push the wear ring off the impeller.
4. These wear rings are exposed to a differential force arising from differential pressures from impeller side to face side, leakage flow. On high-pressure pumps, this delta P causes high axial forces which try to remove the wear ring off the impeller. This force is also proportional to the impeller wear ring thickness height.
5. The rotating wear ring is subjected to rubs with the stationary ring at times such as during startups, and these can cause a high torque on the wear ring and allow it to loosen in combination with the other forces.

Discussion on Reliability Improvements Added to Achieve High Reliability

The reason that the five pumps have a history of excessive failures and maintenance must be one or more of the five reasons below. This is true for any machinery failures. The consultant's studies of the pumps and the failure modes led to an understanding that the majority of problems were due to design errors and incorrect specifications. There were no manufacturing or operational errors.

The Five Root Causes of Machinery Failure

1. **Manufacturer design defect.**
2. **Manufacturing defect: Assembly, machining, casting, welding, material.**
3. **Incorrect specification by customer or installation-system defect**
4. **Incorrect operation**
5. **Incorrect maintenance**

Design Errors

1. The wear rings originally had a medium shrink fit to impellers, plus axial inserted grub screws inserted into threaded holes drilled axially into the wear rings and impeller. The grub screw threaded hole acts as a very high stress raiser in the weak wear rings and in a short time, cracks develop within 9 months maximum, and the rings break off, thus damaging pump. The industry has been using this extremely poor fixation design for many years, and this has caused uncounted pump failures for many years. It is an example of engineering incompetency leading to reliability failures. The first solution proposed by shop personnel and applied in 2007 was to eliminate grub screws and satisfy axial holding power by applying butt welds, each one about 2 cm long, at several locations around wear ring. This was not a good idea; the machines failed after one year service.
2. The second-generation wear ring fixation method only increased lifetime between failures from previous 9 months, to a new MTBO of 12 months. This was because failure analysis of the cracked wear rings found in 2008 showed that the combination of medium shrink fit impellers wear rings with high tensile hoop stresses, plus a weld zone, leads to stress concentration cracks resulting from the weld zone at a thin wear ring, meaning high tensile stresses. In addition, the 15,000 TDS water is corrosive and assists intergranular corrosion process, thus further weakening the assembly. The best new solution ideas developed for the failure modes found were the method of applying Taper fit wear rings with shrink fit and no welding or grub screws, or the full welded version using wearing rings without any shrink fit applied.

3. The wear rings were also exposed to fine sand particles in the water at interface between stationary and rotating wear rings. This led to very high solid particle plus liquid erosion and loss of performance. The OEM should have used a high hardness weld overlay such as Stellite with RC-45 surface hardness and higher. The solution is to apply weld overlay using Stellite welding rods onto the rotating wear ring outer surface and the stationary wear ring inner dimeter. We advised two solution options to the shop: Either use RC-45 hardness Stellite weld overlay or utilize HVOF spray coatings of Tungsten Carbide giving about Rockwell-C 50 hardness. They decided to use HVOF spray which worked well in this case. That was from 2008 to 2010. Later, in 2013, we started using Laser claddings on wear ring "contacting" surfaces with hardness of Rockwell-C 55 and these gave excellent service as sometimes the HVOF coating would flake off partially.

Manufacturing Errors: None Found

User Specification Errors

1. The pumps were shown in customer supplied data sheets as having 15,000 PPM total dissolved salts, but no solids were mentioned. In actuality, the water wells contained a high amount of fine sand that entered the injection pumps leading to erosion in a short time of wear rings, balance piston and its seals, and even casing internals at volutes tips or tongues. This is not the fault of the manufacturer. The user should have requested specific high hardness coatings or weld overlay for any very high power pumps in water services as water is aggressive in terms of erosion at high velocities even without solid particles.

User Maintenance Errors: None Found

From a reliability point of view, the reader can recognize that machinery failures are not the result of one failure mode but multiple failures modes.

These pump failure modes included: a) solid particle erosion on many internal parts, b) wear ring cracking and acting as foreign object damage. The subsequent debris entered the other sections of the pump and then caused high unbalance vibration when broken parts got stuck inside impellers. These failure modes acted together to drastically reduce operating MTBF. The average lifetime of the pump cartridges in this difficult service is now 4.0 years, which is considered acceptable to the end users. With further improvements at the water wells to upgrade the sand trapping capability, the life should extend to 6.0 years. The pumps also experiencing mechanical seal failures, caused by various seal design issues, which were all finally resolved.

About the Editor

Robert Perez is mechanical engineer with more than 40 years of rotating equipment experience in the petrochemical industry. He has worked in petroleum refineries, chemical facilities, and gas processing plants. He earned a BSME degree from Texas A&M University at College Station, an MSME degree from the University of Texas at Austin and holds a Texas PE license. Mr. Perez has written numerous technical articles for magazines and conferences proceedings and has authored 5 books and coauthored 4 books covering machinery reliability. He is also the technical editor of *Kane's Rotating Machinery Dictionary.*

About the Contributors

Mr. LeBleu has taken and taught many classes on simple vibration analysis, lubrication, alignment, reciprocating compressors, centrifugal compressors, fans, centrifugal pumps, positive displacement pumps and mechanical seals and packing as well as set up predictive and preventative maintenance programs. He is familiar with motors, generators and gas and steam turbines as well. His field of experience involves rotating equipment failures of all types such as pumps, motors, the mechanical portion of the failure, seals, bearings, turbines both steam and gas, fans, and couplings. He has over 35 years of experience on rotating equipment of most every kind.

Mr. LeBleu has written numerous procedures for the maintenance, operation, and overhaul of pumps, compressors and rotating equipment. He taught inspection of heat exchangers, columns, vessels, piping and valves to engineers and operators and also taught for two years in power generation at an Air Force technical school.

Drew Troyer, CRE is a Principal with T.A. Cook Consultants. Drew has more than 30 years of experience in the asset reliability management industry. He was a cofounder of Noria Corporation, where he served from 1998–2010.

Drew is a global thought leader in the field of proactive and precision maintenance and was the creator of his popular workshops "Focus on FLAB (fasteners-lubrication-alignment-balance) with Proactive and Precision Maintenance" and "Plant Reliability in Dollars & $ense."

Drew has authored more than 300 books, chapters, articles and technical papers, including co-authorship of the popular book Oil Analysis Basics.

Drew is a Certified Reliability Engineer (CRE) and an MBA. He recently completed a master's degree in environmental sustainability at Harvard University and is very passionate about sustainable manufacturing.

Abdulrahman Alkhowaiter, BSME, is a Rotating Equipment Consultant & Instructor with over thirty two years of experience covering a wide variety of machinery engineering and troubleshooting experience. This experience includes Maintenance Reliability positions covering Reliability Centered Maintenance of compressors, steam turbines, gas turbines, diesel engines, pumps, electric motors, gearboxes, blowers, and fans for seven years. In addition, he has worked 22 years in engineering consulting for a major petroleum company specializing as a compressor, gearboxes, couplings, and steam turbine specialist. Abdulrahman teaches international courses in Machinery Failure Analysis, Process Plant Compressors, Steam Turbines Field Troubleshooting, and Machinery Reliability Improvement.

Index

Also of Interest

Other books by the same editor

DESIGN, MODELING AND RELIABILITY IN ROTATING MACHINERY, By Robert X. Perez, ISBN: 9781119631682. This broad collection of current rotating machinery topics, written by industry experts, is a must-have for rotating equipment engineers, maintenance personnel, students, and anyone else wanting to stay abreast with current rotating machinery concepts and technology. *NOW AVAILABLE!*

Troubleshooting Rotating Machinery: Including Centrifugal Pumps and Compressors, Reciprocating Pumps and Compressors, Fans, Steam Turbines, Electric Motors, and More, by Robert X. Perez and Andrew P. Conkey, ISBN 9781119294139. A must-read for anyone whose job depends on operating or maintaining safe, reliable, and efficient machinery in a process facility, this volume shows the newcomer and veteran alike how you can start troubleshooting rotating equipment problems immediately by implementing this book's common-sense methodology. *NOW AVAILABLE!*

Operator's Guide to Process Compressors, by Robert X. Perez, ISBN 9781119580614. The perfect primer for anyone responsible for operating or maintaining process gas compressors. *NOW AVAILABLE!*

Operator's Guide to General Purpose Steam Turbines: An Overview of Operating Principles, Construction, Best Practices, and Troubleshooting, by Robert X. Perez and David W. Lawhon, ISBN: 9781119294214. A must-read book for anyone operating or maintaining general purpose steam turbines in a process setting, this is a practical reference that includes a collection of field-tested procedures aimed at ensuring safe general purpose steam turbines start-ups, shutdowns, and over-speed tests. *NOW AVAILABLE!*

Check out these other titles from Scrivener Publishing

Dictionary of Industrial Terminology 2nd Edition, by Michael D. Holloway and Emma Holloway, ISBN 9781119363446. Greatly expanded and revised, this second edition is the most comprehensive and in-depth dictionary of industrial terms available, a must-have for any engineer, scientist, technician, or operator working in industry. *NOW AVAILABLE!*

Basic Electrical and Instrumentation Engineering, by P. Sivaraman, C. Sharmeela, A. Thaiyal Nayagi, and R. Mahendran, ISBN 9781119764465. Offering the most up-to-date coverage on the subject available, this new volume provides detailed concepts of AC circuits and power systems, transformers, DC machines, AC machines, measurement and instrumentation, and many other practical concepts for the engineer or engineering student. *NOW AVAILABLE!*

Electrical and Electronic Devices, Circuits, and Materials: Technological Challenges and Solutions, edited by Suman Lata Tripathi, Parvej Ahmad Alvi, and Umashankar Subramaniam, ISBN 9781119750369. Covering every aspect of the design and improvement needed for solid-state electronic devices and circuit and their reliability issues, this new volume also includes overall system design for all kinds of analog and digital applications and developments in power systems. *NOW AVAILABLE!*

Printed in the USA/Agawam, MA
January 9, 2023

804046.001